Scour Manual

Scour Manual

Current-Related Erosion

SECOND EDITION

Edited by

G.J.C.M. Hoffmans
Deltares, Delft, The Netherlands

H.J. Verheij
Deltares, Delft, The Netherlands

CRC Press
Taylor & Francis Group
Boca Raton London New York

CRC Press is an imprint of the
Taylor & Francis Group, an **informa** business

A BALKEMA BOOK

Cover image: depth sounding commissioned by Rijkswaterstaat at the weir at Grave after a collision with a ship.

Published by:
CRC Press/Balkema
Schipholweg 107C, 2316 XC Leiden, The Netherlands

ISBN: 978-0-367-67594-3 (hbk)
ISBN: 978-0-367-67597-4 (pbk)
ISBN: 978-1-003-13196-0 (ebk)

Typeset by codeMantra

Visit the Taylor & Francis Web site at
http://www.taylorandfrancis.com

and the CRC Press Web site at
http://www.crcpress.com

Library of Congress Cataloging-in-Publication Data

DOI: 10.1201/10.1201/b22624
https://doi.org/10.1201/b22624

Editorial Board

Contributors

Contributors 2020

M. van der Wal (Deltares)
E. Mosselman (Deltares)

Contributors cases

M. Kroeders (DEME group)
T. Blokland (Rotterdam Port Authority)
P. Groenewegen (RoyalHaskoningDHV)
G. Smith (Van Oord)
N.A. van der Leer (Witteveen+Bos)
H.G. Tuin (Arcadis)
J.P.F.M. Janssen (Rijkswaterstaat)
J. Kroon (Svasek Hydraulics)
G.J.C.M. Hoffmans (Deltares)

Contributors 1997

W. Leeuwestein
K.W. Pilarczyk
G.J. Schiereck
G.J. Akkerman
W.H.G. Klomp
J. van Duivendijk
R.O.Th. Zijlstra

Contents

Foreword

This manual is an update of the internationally well-known Scour Manual, published in 1997, in which much of the knowledge and experience at that time on scour was captured. Knowledge that was gathered after the February 1953 disaster where many dikes in the south-western provinces of the Netherlands failed during a severe storm surge. As a consequence of this disaster, several hydraulic and soil mechanical issues had to be dealt with in order to be able to draw up appropriate solutions for the breaches in the flood defenses. In part the solutions consisted of repairing the dikes and in part of constructing closure dams in the estuaries. To study the effects of closures, small-scale experiments were carried out to obtain general information about the critical velocity for the stability of stones and concrete blocks, the overlapping of mattresses, the water movement, and the scouring effects downstream of revetments.

Based on a systematic investigation of the time scale for two- and three-dimensional scour in loose sediments, relations were derived for predicting the maximum scour depth as a function of time. In the 1990s, these scour relations were slightly modified and used for the design of the storm surge barrier in the Nieuwe Waterweg near Rotterdam and for the prediction of the scour process downstream of the barrier in the Eastern Scheldt.

Following the publication of the Scour Manual in 1997, more experience is acquired with existing formulas and the knowledge in the field has widened, especially related to turbulence. The original Scour Manual is partly rewritten to capture this information. Moreover, attention is paid to mathematical scour and erosion models, risk assessment and erosion of cohesive sediments. Some applications of this knowledge in projects are described in case studies at the end of the manual. In this update, it was chosen not to address coastal and offshore structures. For wave-induced scour reference is made to the 1997 version of this manual or to other international manuals.

This update of the Scour Manual concerns the scour processes and phenomena taking place near hydraulic structures due to currents. The manual is intended primarily for hydraulic engineers in the field; however, it may also have appeal to researchers in hydraulic engineering. The scour process has still not yet been explained in a generally accepted manner, and therefore it would be only appropriate to keep discussing their mechanism and formulations for the simpler cases.

The updated Scour Manual was prepared by the original authors and supervised by a CUR committee. The update is dedicated to Mr. G. Vergeer, a strong promotor of this update, who passed away in 2018.

I would like to thank all those who contributed their time and knowledge to the update of the Scour Manual and especially the companies that contributed to the cases.

I wish it will offer the practicing engineer again a guideline in the field.

Pieter van Berkum
Head of Hydraulic Engineering section, Rijkswaterstaat

Acknowledgements

For this updated version of the scour manual, we would like to thank all members of the committee and the contributors for their support and especially Erik Mosselman for reviewing the document and Maarten van der Wal and Frans Buschman for leading the project at Deltares.

Also, the committee members and the contributors to the first version of the manual are acknowledged. They helped creating the basis for the update.

List of main symbols

a	factor, $a = \cot \beta_a + \cot \delta$	–
a	deck block height of a bridge	m
A	cross-section	m^2
A_a	cross-section area of abutment	m^2
A_b	two-dimensional blocking area of abutment	m^2
A_f	cross-section area upstream of abutment	m^2
A_r	cross-section area of the river	m^2
B	length of structure (perpendicular to flow direction), pier width	m
b	factor, $b = \cot \gamma_2 - \cot \gamma_1$ (–)	–
b_u	diameter of pipe or thickness of jet at $x = 0$ (m)	m
B	width of flow	m
B	load factor in propeller scour	–
B_c	stability factor in propeller scour equation	–
B_1	width of the river upstream of the constriction	m
B_2	width of the river at the constriction	m
c	parameter $c = 10^3$ N/m^3	N/m^3
c	cohesion	N/m^2
c_a	shape factor of scour hole, $c_a = 22$	–
c_f	resistance coefficient, $c_f = 0.010$ (range 0.005–0.020)	–
c_s	coefficient of Schoklitsch, $c_s = 4.75$ m$^{0.16}$s$^{0.57}$	–
c_v	velocity distribution coefficient, $c_v = 1.0$	–
c_o	coefficient, $c_0 = 0.29$ (sand) to 1.24 (gravel)	–
c_{2H}	dimensionless parameter for 2D-H jets	–
c_{2V}	dimensionless parameter for 2D-V jets	–
c_{3H}	dimensionless parameter for 3D-H jets	–
C	Chézy coefficient	m$^{1/2}$/s
C_f	fatigue rupture strength of clay, $C_f = 0.035 C_o$ N/m^2	N/m^2
C_k	constant that ranges from 0.030 (for scour slopes less steep than IV:3H) to 0.045 (for backward facing step)	–
C_0	cohesion in the Mirtskhoulava formula	N/m^2
d	particle diameter	m
d_a	size of detaching aggregates, $d_a = 0.004$ m	m
d_0	characteristic length, $d_0 = 1/2\, h_d$, h_d is the drop height	m
d_{50}	median particle diameter for which 50% of the mixture is smaller	m
d_{50f}	median filter size	m
d_{90}	particle diameter for which 90% of the mixture is smaller	m
D	height of sill, step height	m
D	thickness of cohesive layer	m
D	jet or pipe diameter	m
D_F	filter thickness	m
D_p	drop height of grade-control structure	m
D_r	relative soil density	–
D_{90*}	dimensionless grain diameter	–

D_*	sedimentological diameter, $d(\Delta g/v^2)^{1/3}$	–
e	actual void ratio, $e = V_v/V_s$	–
e_{max}	maximum void ratio	–
e_{min}	minimum void ratio	–
f_c	friction coefficient, g/C^2	–
f_c	roughness function, C/C_0, $C_0 = 40\,\mathrm{m}^{1/2}/\mathrm{s}$	–
f_u	undrained shear strength	$\mathrm{N/m^2}$
F	fraction of fines of soil smaller than 0.075 mm	–
F_{down}	downward force	N
F_{lift}	lift force	N
Fr	Froude number related to water depth, $Fr = U_0/(gh)^{1/2}$	–
Fr_s	Froude number related to pressure $Fr_s = \dfrac{V_{uc}}{\sqrt{g(h_u - h_b)}}$	–
Fr_I	Froude number just upstream of the hydraulic jump	–
Fr_I	Froude number in the jet, $Fr_I = U_I/(gb_u)^{1/2}$	–
Fr_I	Froude number, $Fr_I = U_I/(gy_I)^{0.5}$	–
g	acceleration due to gravity, $g = 9.78$–$9.83\,\mathrm{m/s^2}$	$\mathrm{m/s^2}$
G_B	width of scour hole at broken pipeline	m
G_L	length of scour hole at broken pipeline	m
h	flow depth	m
h_b	water depth under a bridge	m
h_e	critical water depth	m
h_c	equilibrium water depth after functioning of a falling apron	m
h_i	stages in the water depth during functioning of a falling apron	m
h_p	distance between propeller axis and bed	m
h_t	tailwater depth	m
h_u	upstream water depth	m
h_0	initial or average flow depth	m
$h_0(0)$	tide-averaged flow depth	m
h_I	average depth in contracted area	m
h_2	average depth in upstream section	m
H	height between head and tailwater levels	m
H	difference in height between upstream and downstream water levels	m
I	volume of scour hole per unit width	$\mathrm{m^2}$
k	turbulent kinetic energy	$\mathrm{m^2/s^2}$
k_m	mean turbulent kinetic energy where scour depth is at maximum	$\mathrm{m^2/s^2}$
k_s	Nikuradse bed roughness (rough: $k_s = 3d_{90}$, smooth: $k_s = 2d_{50}$)	m
k_{max}	maximum turbulent kinetic energy in mixing layer	$\mathrm{m^2/s^2}$
K	non-dimensionless constant, $K = 330\,\mathrm{m^{2.3}/s^{3.3}}$	$\mathrm{m^{2.3}/s^{3.3}}$
K	factor for various influences, such as pier shape and flow angle	–
K_b	coefficient	–
K_I	coefficient, $K/(g^{1.43}\mu^{0.43})$ (K in $\mathrm{m^{2.3}/s^{3.3}}$)	–
I	length of structure parallel to the flow direction	m
L	length of bed protection	m
L_{ins}	bed protection length to prevent shear failure	m
L_{min}	minimum bed length	m
L_p	bridge pier length	m
L_r	length hydraulic jump	m
L_s	length of scour hole	m
L_s	failure length	m
L_s	ship length	m
LL	liquid limit	–
m	constriction ratio, $m = 1 - B_2/B_1$	–
n	porosity, default 0.4	–
n	scale ratio	–
n	Manning's coefficient	$\mathrm{s/m^{1/3}}$
N	number of ship passages	–
$p(\xi')$	probability density function	–

$P(\xi')$	cumulative density function	–
P_f	failure probability	–
PI	plasticity index	–
q	discharge per unit width	m^2/s
q	discharge in a 2D-H jet	m^2/s
q_s	(reduced) sediment transport per unit width (including porosity)	m^2/s
Q	discharge	m^3/s
Q	discharge in a 3D-H jet	m^3/s
Q_c	discharge through main river (without floodplain), see Fig. 3.1	m^3/s
Q_f	discharge floodplain	m^3/s
Q_1	discharge in the upstream channel	m^3/s
Q_2	discharge in the contracted section	m^3/s
r	local turbulence intensity	–
r	discrepancy ratio	–
r_0	depth-averaged relative turbulence intensity, σ_u/U	–
$r_{0,m}$	depth averaged relative turbulence intensity when scour depth is maximal	–
R	strength component	Var.
R	hydraulic radius	m
R	radius of curvature of the centreline	m
R	erosion parameter	$kg/(m \cdot s^3)$
R	erosion rate	m/s
Re	Reynolds number, $Re = Uh/v$	–
$Re*$	Reynolds stress number related to particle diameter, $Re = U_*D/v$	–
s	specific density of bottom material	–
s_b	bed load	m^2/s
s_s	suspended load	m^2/s
S	load components	Var.
S_1	slope energy grade line	–
t	time	s
t_c	time referring to conditions where $q_s = 0$	s
t_p	time referring to live bed conditions	s
t_1	characteristic time at which the maximum scour depth equals h_0	s
t_1	characteristic time at which $y_m = \lambda(s)$	s
$t_{1,u}$	characteristic time at which $y_m = h_0(0)$	s
t_1	time at which αU_0 first exceeds U_c during flood tide	s
t_2	time at which αU_0 drops below U_c during ebb tide	s
T	half tidal period where $\alpha U_0 > U_c$ (s), $T = t_2 - t_1$	s
$T_{0,s}$	time scale for change of the cross-section profile	s
u	local longitudinal flow velocity	m/s
u	mean velocity in the x-direction	m/s
u_m	maximum velocity of u at any x-section	m/s
u_*	bed shear velocity	m/s
$u_{*,c}$	critical bed shear velocity	m/s
U	time-averaged velocity	m/s
U_b	water velocity just above the bed due to the return current	m/s
U_c	critical averaged flow velocity for uniform flow; U_c can be depth-averaged or the near-bed critical velocity	m/s
U_c	critical depth-averaged flow velocity	m/s
U_d	characteristic tidal mean flow velocity	m/s
U_g	mean gap velocity	m/s
U_h	horizontal component jet velocity	m/s
U_m	depth-averaged velocity where scour depth is at maximum	m/s
$U_{m,t}$	maximum velocity during a tide	m/s
U_{max}	maximum velocity	m/s
U_r	ship-induced return current below the ship's keel	m/s
U_v	vertical component jet velocity	m/s
U_0	depth-width-averaged flow velocity, Q/A	m/s

U_0	depth-averaged flow velocity upstream of scour hole	m/s
$U_{0,c}$	critical flow velocity	m/s
U_1	average flow velocity in the jet	m/s
U_1	jet velocity entering tailwater, $U_1 = \sqrt{2gH}$	m/s
U_1	efflux velocity at $x = 0$ (m/s)	m/s
U_2	average flow velocity downstream of the scour hole	m/s
V_s	ship speed (m/s)	m/s
V_s	volume of solids	m3
V_{uc}	critical velocity pressure scour	m/s
V_{ue}	effective velocity pressure scour	m/s
V_v	volume of voids	m^3
$V(t)$	volume of scour hole per unit width	m^3/m
$Vr(t)$	reduced volume of scour hole per unit width	m^3/m
w	water content	–
w_s	fall velocity	m/s
W_1	bottom width in the of the upstream main channel	m
W_2	bottom width in the contracted section	m
x	longitudinal distance	m
y	vertical distance	m
y_c	critical scour depth	m
y_d	depth after a sliding	m
y_m	maximum scour depth	m
$y_m(t)$	maximum scour depth as a function of time	m
$y_{m,e}$	equilibrium scour depth	m
$y_{m,e,max}$	maximum equilibrium scour depth	m
$y_{m,e|50\%}$	equilibrium scour depth exceeded by 50% of the scour	m
y_{ad}	bed elevation changes due to long-term deposition or bed erosion	m
y_{be}	bend scour	m
y_{bf}	bed form trough depth	m
y_{cf}	confluence scour	m
y_{cs}	constriction scour	m
y_s	local scour	m
$y_{s,actual}$	actual scour depth	m
y_{ss}	equilibrium scour depth (static scouring)	m
y_{tot}	total scour depth	m
y_1	thickness of the jet at the vena contracta	m
z	vertical distance from the axis of the jet at any section	m
Z	reliability parameter	Var.
Z	scour number	–
α	flow and turbulence coefficient (or angle)	–
α_F	turbulence coefficient, $\alpha_F = (1 + 3r_0)c_v$	–
α_g	gap coefficient, $\alpha_g = 2.4$	–
α_{ga}	abutment coefficient, $\alpha_{ga} = 1.0$ or 1.4	–
α_{RAJ}	constant, $\alpha_{RAJ} = 0.3$	–
α_1	turbulence coefficient, $\alpha_1 = 1.5 + 5r_0$	–
α_u	coefficient, to be determined as $\alpha_u = \alpha - U_c/U_0$	–
α_v	Veronese coefficient, $\alpha_v = 1.9$	–
α_1	angle upstream wing wall of abutment	°
α_2	angle downstream wing wall of abutment	°
β	upstream scour slope (or angle)	°
β	reliability index	–
β	coefficient; $\beta = 0.67–0.8$	–
β_a	average slope angle before instability	°
γ	coefficient (or angle)	–
γ	weight per unit volume, $\gamma = 10$ kN/m³	kN/m^3
γ	coefficient, $\gamma = 0.4–0.8$	-
γ	coefficient, $\gamma = 0.22–0.23$	-
γ_s	safety factor	-
γ_{wet}	wet weight per unit volume	kN/m^3

γ_l	sliding erosion slope angle after instability	°
γ_2	sliding deposit slope angle after instability	°
δ	slope angle downstream of the point of reattachment	°–
δ	slope angle downstream of the deepest point of the scour hole	°
Δ	relative density, $\rho_s/\rho - 1$	–
Δ_t	time step	s
η	$\eta = z/x$	–
η	coefficient, $\eta = 0.75{-}0.85$	–
η_b	coefficient	–
η_s	coefficient	–
θ	temperature	°C
θ	angle between two upstream branches of a confluence	°
θ	jet angle near surface	°
κ	constant of von Kármán, $\kappa = 0.4$	–
λ	characteristic length scale	m
μ	shape factor or roughness factor	–
μ	average value	Var.
μ_R	average value of the strength parameter R	Var.
μ_S	average value of the load parameter S	Var.
μ_Z	average value of the reliability parameter	Var.
v	kinematic viscosity	m^2/s
ξ	ratio measured and calculated scour depth	–
ρ	fluid density	kg/m^3
ρ_b	density bed material	kg/m^3
ρ_s	material density	kg/m^3
σ	relative standard deviation	–
σ_g	sediment gradation, $\sigma_g = d84/d50$	–
σ_u	standard deviation of the instantaneous longitudinal velocity averaged over the depth	m/s
σ_v	standard deviation of instantaneous velocity in transverse direction	m/s
σ_w	standard deviation of instantaneous velocity in vertical direction	m/s
σ_Z	standard deviation of the reliability parameter z	Var.
τ_c	critical bed shear stress	kg/m.s^2
τ_0	bed shear stress	kg/m.s^2
φ	dimensionless transport rate	–
ϕ	angle of repose, $\phi = 40°$	°
φ'	angle of internal friction	–
χ_e	turbulence parameter	–
ψ	mobility or Shields parameter	–
ψ_c	critical Shields parameter	–
ω	angle of attack	°
ω	turbulence coefficient, $\omega = 1 + 3r_0$	–
ω	fall velocity	m/s
Ω	current-related sediment mobility, $= \psi/\psi_c$	–

List of main definitions

- General scour: degradation of the main channel bed due to an imbalance of the sediment transport entering and leaving a control volume. This occurs, for example, when the sediment transport capacity increases due to accelerating flow.
- Bend scour: local scour in the outer bend of a river due to helical flow; maximum scour depth usually at the downstream side of the bend.
- Constriction scour: local scour due to the transition into a narrower or shallower section of the river.
- Confluence scour: scour due to the confluence of flows from two upstream river branches.
- Non-uniform flow: flow where flow velocity and other hydrodynamic phenomena differ spatially.
- Equilibrium scour depth: constant scour depth reached after some time, when the conditions remain constant.
- Plunging jets: water jets that fall from a certain height onto a free water surface.
- Submerged jets: water jets with an outflow opening under water.
- Grade control structure: structure to control the water level where water flows over or through the structure.
- Shear failure: soil mechanical instability of non-cohesive soil.
- Flow slide: soil mechanical instability of loosely packed sand.
- Two-dimensional scour: scour as a result of a long (normal to the flow direction) sill or infinitely long outflow opening. The scour is equal over the full length or width. A typical example is an underwater sill in a closure of an estuary.
- Three-dimensional scour: scour downstream of a structure with a limited width (normal to the flow direction). At the sides of the outflow opening, eddies occur causing extra scour. Typical examples: spurs, outflow of culverts.
- Clear water scour: scour with transport of bed material only due to the presence of a structure. Without the structure, no bed load or suspended sediment transport would occur. This type of scour occurs in laboratory conditions and in areas with scour-resistant beds. It results in deeper scour holes.
- Live-bed scour: scour with supply of bed material from upstream.
- Bed load transport: sediment transport by rolling, sliding and saltating (jumping up into the flow, being transported a short distance then settling) of sediment particles mainly just above the river bed.

- Suspended load transport: transport of sediment that is suspended in the water column by turbulence. The suspended load usually consists of smaller sediment particles than bed load.
- Abutment: horizontal construction into the flow as part of an approach embankment for a bridge.
- Groyne or spur: horizontal constriction of a flow to train a river to provide sufficient depth for navigation and to prevent erosion of the river bank.

Chapter 1

Introduction

1.1 General

A hydraulic structure is generally intended to provide a practical measure to solve an identified problem. After problem identification, subsequent stages are determined by a series of decisions and actions culminating in the creation of a structure or structures to resolve the problem. Aspects that may affect the eventual outcome of the design process have to be assessed. In addition to hydraulic, geotechnical and engineering characteristics, aspects such as social conditions, economics, environmental impact and safety requirements also influence the design process.

Within the scope of the Dutch Delta works, the Dutch Ministry of Transport, Public Works and Water Management and Delft Hydraulics (now merged into Deltares) conducted systematic research with respect to the prediction of the formation of scour holes. After the catastrophic flood disaster in 1953, the Delta Plan was formulated to protect the Rhine-Meuse-Scheldt delta against future disasters. Dams with large-scale sluices were planned in some estuaries. The expected severe scour necessitated acquiring a better understanding of the scour process.

To obtain detailed information about the physical processes playing a role in scour development, Delft Hydraulics (Deltares) carried out many experiments in which several parameters of the flow and the scoured material varied. From the results of experiments in flumes, with obvious difficulties of scale effects and limitations in instrumentation, some semi-empirical relations were obtained that describe the erosion process as a function of time and position (Breusers, 1966, 1967; van der Meulen & Vinjé, 1975). In addition, design criteria were deduced for the length of bed protection. These were based on hundreds of shear failures and flow slides that occurred along the coastline in the south-western part of the Netherlands.

Understanding of the physical processes and mathematical modelling of the water and sediment movement in rivers, estuaries and coastal waters have made much progress in recent years. This has led to a number of more or less ready-to-use mathematical model systems, but it has also raised many new research questions. In the early 1990s, a morphological model for the generation of scour holes behind hydraulic structures was developed. This morphological model was based on the 2D Navier-Stokes and convection-diffusion equations and used for the calibration and verification of semi-empirical relations that predict the scour process. Nowadays sophisticated CFD (Computational Fluid Dynamic) models are available for scour computations.

This manual highlights the so-called Breusers method which describes the maximum scour depth as a function of time, including the practical equilibrium value near hydraulic structures. Scour due to three-dimensional flow can easily be predicted when this method is applied in combination with computational results of depth-averaged hydrodynamic models or with measurements obtained from scale models. The accuracy of the scour computation depends mainly on the accuracy of the flow velocities and the turbulence intensities just above the protected bed. According to Breusers (1966), the development of the scour process depends entirely on the average flow velocity and depth-averaged relative turbulence intensity at the transition from the fixed to the erodible bed. Applying this concept restricts the scour prediction to a single computation. No information is needed concerning the near-bed velocities and bed turbulence in the scour hole.

This manual is an update of the original book published in 1997. However, it deals only with scour due to currents. Scour due to waves is not addressed in this update. This manual addresses various new aspects, such as risk assessment, scour of rock and new theory-based formulas for the prediction of scour at hydraulic structures but also an update of the available mathematical scour and erosion models. Last but not least, a fully renewed chapter has been added with recent experiences of consultants and contractors with scour design.

1.2 Scope of this manual

The purpose of this scour manual is to provide the civil engineer with useful practical methods to calculate the dimensions of scour holes in the prefeasibility and preliminary stages of a project, and to furnish an introduction to the most relevant literature. The manual contains guidelines which can be used to solve problems related to scour in engineering practice and also reflects the main results of all research projects in the Netherlands in recent decades. A complete review of all the available references on scour is beyond the scope of this manual. The most relevant manuals are Breusers and Raudkivi (1991), Melville and Coleman (2000) and May et al. (2002). Furthermore, the International Conference on Scour and Erosion provides relevant information and is important for the most recent developments.

The scour depth as a function of time can be predicted by the so-called Breusers equilibrium method. Basically, this method can be applied to all situations where local scour is expected. However, the available knowledge about scour is not sufficient for applying the method to scour at each type of structure. Structure-specific scour prediction rules are presented then. The treatment of local scour is classified according to different types of structures. Each type of structure is necessarily schematised to a simple, basic layout. The main parameters of a structure and the main parts of the flow pattern near a structure are described briefly insofar as they are relevant to the description of scour phenomena. Detailed and theoretical descriptions of the flow phenomena are not included because at this stage, the consequences of such descriptions are minimal in relation to engineering practice. Nonetheless, Hoffmans (2012) developed new formulas for equilibrium scour. Evaluating a balance of forces for a control volume, he was able to develop scour equations for different types of flow fields and structures, i.e. jets, abutments and bridge piers.

As many scour problems are still not fully understood, attention is paid to the validity ranges and limitations of the formulas, as well as to the accuracy of calculations of the maximum scour depth during the lifetime, the upstream scour slope and the failure length. Due to shear failures or flow slides, the scour process can progressively damage the bed protection. This will lead to the failure of hydraulic structures.

The presented Breusers equilibrium method can be applied directly in engineering practice for nearly all types of structures. Accurate local flow velocities and turbulence intensities resulting from three-dimensional flow models can act as inputs for the Breusers equilibrium method, which can be considered as a continuation and an expansion of the work of Breusers (1966). In other words, one may speak of a revitalisation of the Breusers formula, with which a lot of experience has been gained, mainly in the Netherlands but also abroad.

1.3 Reading guide

The manual is divided into seven parts. The first three parts give a general introduction to the subject. The next four parts deal with calculation methods for predicting scour near hydraulic structures and, in the final part, some cases of scour at prototype scale are described. A brief summary of each chapter follows.

Chapter 2 – Design process

It is crucial to design hydraulic structures that are reliable and safe during their life cycle. To ensure safe long-term functioning of hydraulic structures, it is necessary to consider boundary conditions, risk assessment and measures to prevent scour. After having addressed the boundary conditions, we discuss the risk assessment and the fault tree analysis. Two methods are treated: one based on safety factors and the other on failure probability. When applying these techniques, one should keep in mind what the goal is of the design: a pre-feasibility study or a final design. Examples show how to deal with these methods. Furthermore, protective measures are mentioned.

Chapter 3 – Design tools

The total scour which may occur at the site of a structure can be estimated with mathematical scour and erosion models. An overview is given of available tools. In principle, scour may be considered as a combination of general scour and local scour resulting from different processes. In addition, time phases can be distinguished in the scour process. We present these phenomena for currents.

A more or less continuous scouring process may suddenly be disturbed by the occurrence of geotechnical instabilities along the upstream scour slope. Shear failures and flow slides influence the stability of hydraulic structures. In the extreme case, these instabilities involve large masses of sediment and cause a major change of the shape of the upstream side of the scour hole in a relatively short period of time. Some design criteria based on storage models are presented.

Chapter 4 – Initiation of motion

Scour results from transport of bed materials. The non-uniform flow is responsible for this, which is usually expressed by either a turbulence coefficient or a dominating flow velocity, or by both. Turbulence is the most important phenomenon determining erosion. Relations for the turbulence intensity and the critical flow velocity are presented for various situations. The design graphs of Shields are presented for non-cohesive bed materials, such as sand and rock. For cohesive soils such as clay and peat, the method of Mirtskhoulava and also empirical relationships based on the plasticity index are given. In addition, erosion rate formulas are also presented.

Chapter 5 – Jets

We discuss scour due to several jet forms, such as plunging jets, submerged jets, horizontal and vertical jets, and two- and three-dimensional jets. In addition, we treat the complex flow pattern of jets. We also address scour by ship-induced currents, scour due to propellers and scour due to jets in the case of broken pipelines. Semi-empirical and theory-based relations for the scour process behind a short-crested sill are presented. The semi-empirical relations are often used for grade-control structures, where the flow above the structure is supercritical, and for the time-dependent development of the maximum scour depth downstream of a hydraulic jump. The structure of the semi-empirical relations shows a good similarity with the Breusers approach. The new relations have a theoretical base as they have been derived using the balance of forces. However, both semi-empirical and theory-based relations must be clearly understood prior to any attempt to use them for design purposes.

Chapter 6 – Sills

We summarise calculation methods for sills. A distinction is made between sills with a broad or a sharp crest and between sills with and without bed protection. Usually, the flow above a sill is subcritical, but depending on the downstream water level, the flow may become supercritical. We discuss the time-dependent and equilibrium behaviour of scour holes in sandy beds in relation to closure works (broad-crested sills) in tidal channels. Special attention is paid to the effects of turbulence and flow pattern on the scour process.

We describe an approximate method (reduction method) for calculation of the maximum scour depth. This takes the influence of upstream sediment supply into account. In addition, we present a method to adjust this calculation method for unsteady flow, especially tidal flow. These methods were successfully applied during the design of the Eastern Scheldt Storm Surge Barrier. The upstream scour slope determines the stability of the upstream part of the scour hole and the adjacent bed protection. A relation for the upstream scour slope, based on a probabilistic model for bed load transport, is presented. Relations derived from systematic scour investigations are verified by two field experiments, among which the scour at the Eastern Scheldt Storm Surge Barrier.

Chapter 7 – Abutments and groynes

Relations are presented for predicting local scour at the head of abutments, for which several names are used (spurs, groynes, guide or river bunds) in the literature. We also present recently developed formulas based on a balance of forces. We briefly discuss the flow characteristics around blunt and streamlined abutments. Attention is also paid to the time scale of the scour process and to combined scour (e.g. local scour and bend scour or constriction scour). Since the literature contains many scour relations, a number of generally acceptable predictors have been selected for this manual. Finally, attention is paid to failure mechanisms and measures to mitigate scour near abutments.

Chapter 8 – Bridge piers

After discussing the flow characteristics at the pier and in case of a submerged bridge, relations for estimating scour around bridge piers are summarised. These relations are mostly empirical, but we also present a theoretical relation based on a balance of forces. Correction factors and design graphs for the equilibrium scour depth are discussed. Attention is paid both to the equilibrium scour depth and to the time scale of the scour process. Methods are given to predict scour at bridge piers with a footing or pile cap and for pressure scour. Indications are provided for determining the area to protect against scour.

Chapter 9 – Realised case studies on prototype scale

Nine realised case studies on the prototype scale, based on feasibility studies or design studies, are evaluated in order to determine the practical use of the scour relations in this manual. These cases are as follows:

- Four cases about bridge pier scour: Camden motorway bypass, crossing of a high-voltage power line, pier scour in bypass channel, and pressure scour;
- Two cases about culvert scour: Waterdunen project, and scour development in front of a culvert;
- Two cases about jet scour: propeller- and thruster-induced jet scour;
- One case about sill scour: weir at Grave.

Chapter 2

Design process

2.1 Introduction

Hydraulic structures cause disturbances in uniform flow and sediment transport. Downstream of these structures, flow velocities increase due to constriction by the structure. When the flow velocities decrease (i.e. in the deceleration zone), a higher degree of turbulence is generated. Therefore, a stronger erosion capacity is present. In most cases, this leads to scouring and, depending on the specific hydraulic conditions, there are sometimes steep upstream slopes of scour holes. The bed is often protected in order to decrease the maximum dimensions of the scour hole and to shift the scour holes that involve a potential risk to structural stability to a greater distance from the hydraulic structure. The main dimensions of the scour hole can be characterised roughly by the maximum scour depth expected during the lifetime of the structure and by the upstream scour slope. Both hydraulic and geotechnical characteristics influence these two design parameters and are treated in more detail in subsequent sub-sections.

To ensure the safety and long-term functioning of hydraulic structures, it is necessary to consider information with respect to failure mechanisms, boundary conditions and design criteria. In recent years, the need for reliable modelling of sediments exposed to wave and current action has been increased. This need arises partly from increasingly high safety standards. When designing structures, the following aspects must be considered (Pilarczyk, 1995):

- Function of the structure: erosion as such is not the problem as long as the structure can fulfil its function.
- Physical environment: the structure should offer the required degree of protection against hydraulic loading, with an acceptable risk and, when possible, meet the requirements resulting from landscape, recreational and ecological viewpoints.
- Construction method: the construction costs should be minimised to an acceptable level and legal restrictions must be adhered to.
- Operation and maintenance: it must be possible and feasible to manage and maintain the hydraulic structure.

Elaboration of these points depends on specific, local circumstances, including the type of terrain (lowland or highland) and its development (economic value) and availability of equipment, labour and materials. These circumstances constitute the boundary conditions for design (Section 2.2). The cost of construction and maintenance is

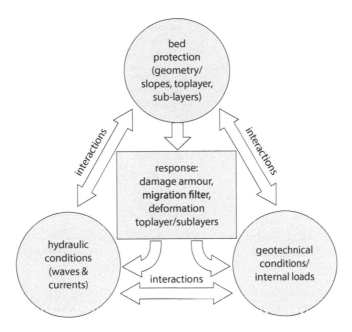

Figure 2.1 Soil–water–structure interaction, e.g. SOWAS concept (de Groot et al., 1988).

generally a controlling factor in determining the type of structure to be used. The starting points for the design should therefore be carefully examined in co-operation with the future manager of the project. Most research problems concerning water defences are multidisciplinary in character, especially in the technical sense, as characterised by all relevant interactions between the soil, water and structure, and may lead to combined hydraulic, geotechnical and structural research. These interactions may be brought together in the diagram shown in Figure 2.1 or in a fault tree (Section 2.3.2).

An example of the design process for a bed protection in relation to expected (predicted) local scour is illustrated in Figure 2.2. In subsequent sections, some design aspects are discussed in greater detail (Section 2.4).

During the design process of hydraulic structures, various stages can be recognised:

1. Pre-feasibility design (often the tender phase)
2. Preliminary design (often the tender phase)
3. Final design or executional design (execution stage).

The first two stages require a rough estimate of the dimensions of the scour holes. A better estimate has to be made in the final design. In the pre-feasibility and the preliminary stages, nearly all presented equations in this manual can be applied. Very often the upper limit will be used because time is limited and not all aspects can be studied in detail. In the final stage, it is recommended to apply a risk-based approach (Section 2.3). Examples of how to deal with failure probability or safety factors are discussed in Section 2.5. Sometimes, during executional stages of large or complex hydraulic structures, experimental research will be necessary. For example, extensive

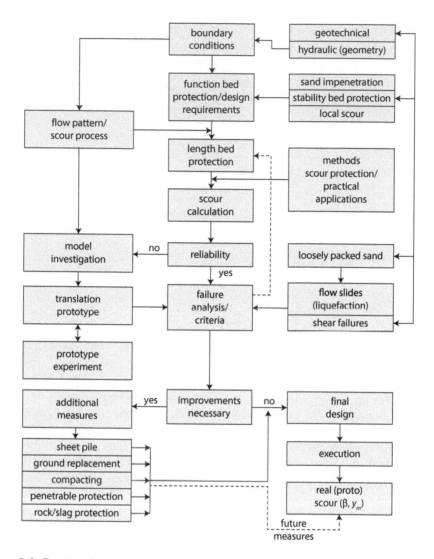

Figure 2.2 **Design phases of the bed protection.**

physical model studies were carried out for the Eastern Scheldt Storm Surge Barrier. Shortly, protective measures to prevent scour will be discussed in Section 2.4.

2.2 Boundary conditions

2.2.1 Introduction

In order to design hydraulic structures, loading (hydraulic conditions) and strength (morphological and geotechnical conditions) parameters have to be specified. Flow characteristics (flow velocities, water levels and discharges) and turbulence intensities determined by the geometry of the hydraulic structure and bed roughness characterise the flow pattern in the vicinity of the hydraulic structure and thus provide a measure of the erosion in the scour hole just downstream of the hydraulic structure. On the other hand, the scour process is also determined by the composition of the bed material (subsoil).

2.2.2 Hydraulic conditions

The most-simple flow pattern is generated by a steady, uniform flow. However, special conditions for the flow pattern can be distinguished, for example an accelerated flow in a local constriction, a river bend with the well-developed bend flow, an unsteady flow due to flood waves or tidal movement, or a flow downstream of hydraulic structures with a direction perpendicular or inclined to the flow direction of a receiving river or estuary. An ice cover can divert the flow to the part of the bed near the hydraulic structure, resulting in an additional increase of local scour. The main hydraulic and special conditions are summarised in Table 2.1.

Table 2.1 Boundary conditions

Loading	Hydraulic conditions	Mean and local flow velocities (water levels and discharges)
		Turbulent energy or turbulence intensity (geometry of structure)
		Kinematic viscosity (influence of water temperature). Fluid density
	Special conditions	Supercritical/subcritical flow
		Unsteady/steady flow
		Ice cover
		Earthquakes
Strength	Morphological conditions	Grain size distribution (representative diameter). Material density
		Non-cohesive/cohesive sediments
		Additional properties of non-cohesive sediments:
		– Shape of grains
		– Surface packing of grains (homogeneity of bed material)
		– Multiple layers of different bed materials
		Additional properties of cohesive sediments:
		– Sodium Adsorption Ratio (SAR)
		– Cation Exchange Capacity (CEC)
		– pH level of pore water
		– Temperature
		– Organic content
		– Porosity
	Geotechnical conditions	Stratification of subsoil
		Bulk density of sand (loosely or densely packed)
		Angle of internal friction
		Porosity
		Cohesion

2.2.3 Morphological conditions

For convenience, the sediments forming the boundaries of a flow are subdivided into cohesive and non-cohesive sediments, although there is a fairly broad transition range. In non-cohesive sediments such as sands and gravels, the particle or grain size and material density are the dominant material parameters for sediment transport. Widely graded bed materials will be more resistant to scour than uniform bed materials of the same median grain size. During a flood, the finer grains of a non-uniform bed material are eroded preferentially, increasing the median grain size of the remaining bed material. This process is known as winnowing. The resulting coarsening of the river bed can ultimately lead to armouring.

The shape of grains, the surface packing of grains and multiple layers of different bed materials are additional factors for the scour process. They are of secondary importance and therefore not treated in detail in this manual. The physio-chemical properties of cohesive sediments play a significant role in the resistance of cohesive sediments against currents. These properties depend strongly on granulometric, mineralogical and chemical characteristics of the sediment involved (Table 2.1 and Section 4.4).

2.2.4 Geotechnical conditions

A purely hydraulic and morphological approach to a given geometry (structure, bed protection) and hydraulic boundary conditions leads to scouring in which the maximum scour depth gradually increases and the upstream scour slope steepens (at least the steep part will become longer) until it reaches the equilibrium phase. This more or less continuous process may suddenly be disturbed by the occurrence of geotechnical instabilities along the upstream scour slope. In the extreme case, these instabilities involve large masses of sediment and cause a major change of the shape of the upstream side of the scour hole in a relatively short period of time. The steeper this slope, the greater the probability of slope failure. Although of minor importance, the maximum scour depth also plays a role.

Besides these geotechnical aspects, the soil properties are important, especially with regard to the type of geotechnical instability that may occur. Two types of instability are distinguished for cohesionless sediment, namely shear failure and flow slide. To predict the occurrence of a shear failure, the steepness of the upstream slope has to be assessed in relation to the angle of internal friction of the bed material. A flow slide is a more complex geotechnical phenomenon which can only occur in loose to very loose sand. However, the final geometrical characteristics of the upstream slope are generally of much greater importance in relation to flow slide instability than they are for a shear failure. The main geotechnical parameters are presented in Table 2.1. Section 3.5 gives more information about geotechnical instabilities.

2.3 Risk assessment

2.3.1 Introduction

The aim of risk assessment is to establish the risks that need to be managed and to identify means to control them to acceptable levels. The risk assessment process should identify the hazards, along with the events or circumstances that may produce them, determine the risk posed by them, and identify the measures that can be put in place to control the risk by preventing the hazard or mitigating its effect if it occurs. In the context of this manual,

- Hazard is defined as conditions with the potential to cause scour.
- Risk is defined as the combination or product of frequency of occurrence and consequence of a hazard.

Prior to making a risk assessment, a safety philosophy needs to be chosen. This manual presents for various structures (sluices, sills, bridge piers and abutments) formulas to estimate the time-dependent scour process including the equilibrium scour depth. Sometimes, these formulas are best-guess predictors; sometimes they give an upper limit. A considerable improvement could be obtained if the formulas are based on a safety philosophy. An example of a risk assessment, including a safety philosophy, is presented by Johnson (1992; Table 2.2), showing for bridge piers a safety factor as a function of the failure probability based on Monte Carlo simulations.

A safety philosophy for scour issues has been followed in a guideline for structures (TAW, 2003) in the Eurocodes NEN-EN-1991, that specify how structures should be designed within the European Union, and in tools for assessing the dike safety used in the Netherlands.

In principle, two approaches can be used for the safety philosophy:

1. Using a "fixed" safety factor, for example a value of 2 for all types of structures, to be defined preferably using a failure requirement.
2. Applying a (semi-)probabilistic approach by deriving a safety factor (or partial safety factors) as a function of the failure probability of a particular formula on the basis of probabilistic computations for a specific structure. Two options are possible: (2a) a fully probabilistic approach or (2b) derivation of a semi-probabilistic rule using probabilistic analyses. Both options are summarised below as probabilistic approaches.

Obviously, approach 2 is preferred over approach 1. However, approach 1 is simpler and should be considered as a fall-back approach. Table 2.3 shows the advantages and disadvantages of approaches 1 and 2.

Next, some theoretical aspects of a failure probability approach are discussed as well as ways to determine the failure probability given uncertainties in relevant parameters. In addition, the safety requirements of the Eurocodes are applied because structures can lose stability due to scour. Exceptions are the Dutch storm surge barriers. These primary water defences have to fulfil the safety requirements of the Dutch Water Law.

Table 2.2 **Relation between safety factor and probability of failure for bridge piers (Johnson, 1992)**

Probability of failure	0.1	0.01	0.001	10^{-4}	10^{-5}
Safety factor	1.2	1.4	1.6	1.75	1.85

Table 2.3 Comparison between a probabilistic approach and an approach based on a safety factor

	Probabilistic analyses	*Fixed safety factor*
Advantages	Determination of safety factor as a function of the failure probability Partial safety factors can be determined All uncertainties can be considered individually	Simple Quick application
Disadvantages	Time consuming Distributions of all parameters should be available	Does not give a realistic idea of the safety No general accepted values for a subjective safety factor, as for example a value of 2
Application stage	Final design	Tender stage Small projects

2.3.2 Fault tree analysis

The design process is characterised by solving design problems in an iterative manner. Since the processes involved are dynamic, it is impossible to reach the optimum solution at the first attempt. Though the optimum solution will never be attained, the design philosophy which has been adopted helps to prevent a haphazard approach to design and research. Figure 2.3 gives a general overview of the failure mechanisms of an open bed protection (Vrijling, 1990).

To produce a safe and reliable design, the total reliability as a function of all modes of failure should be approximated, at least at a conceptual level. For example, Figure 2.4 shows the fault tree for a bed protection in which the foundation of the hydraulic structure is the central point. The bed protection has to prevent or slow down any change in the bed protection and thus in the geometry of the foundation. A failure of the bed protection does not directly imply the loss of the structure, but the resistance of the foundation is reduced if the subsoil becomes unstable owing to the presence of a well-developed scour hole.

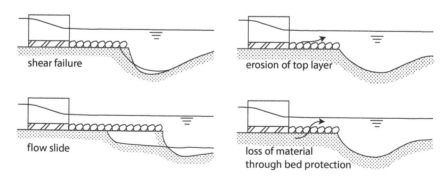

Figure 2.3 Failure modes of an open bed protection.

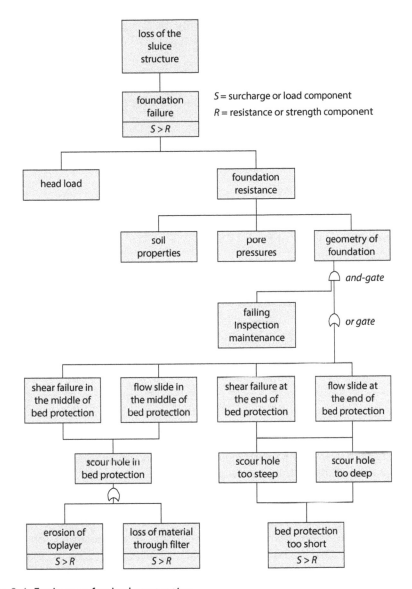

Figure 2.4 **Fault tree for bed protection.**

A further advantage of fault-tree analysis is that this makes it possible to incorporate the failure of mechanical or electrical components as well as human errors in the management and maintenance of the structure. For instance, the safety of a sluice can be dramatically improved by regular echo-sounding of the bed protection and by subsequent maintenance if the initiation of a scour hole is discovered. The probability of instabilities affecting the foundation is thus only high if both a scour hole is formed and inspection and maintenance of the bed protection fail.

More information about the design process, including outlines of main considerations relating to deterministic and probabilistic design processes, can be found, for example in the Rock Manual (CIRIA/CUR/CETMEF, 2007).

2.3.3 Safety factor

The total scour depth needed for a design may be computed by summing all the components of vertical bed change:

$$y_{tot} = y_{ad} + y_{cs} + y_{be} + y_{cf} + y_{bf} + y_s \tag{2.1}$$

where

y_{tot} = total scour depth (m)
y_{ad} = bed elevation changes due to long-term sediment deposition (aggradation) or bed erosion (degradation) (m)
y_{cs} = constriction scour (m)
y_{be} = bend scour (m)
y_{cf} = confluence scour (m)
y_{bf} = bed form trough depth, accounting for the effect of variation due to bedforms (m)
y_s = local scour depth associated with a structure (m)

Applying the method with a safety factor γ_s, Equation (2.1) can be rewritten into

$$y_{tot,\max} = \gamma_s \left(y_{ad} + y_{cs} + y_{be} + y_{cf} + y_{bf} + y_s \right) \tag{2.2}$$

As mentioned before, Johnson (1992) developed a probabilistic approach for pier scour engineering using Monte Carlo simulation. He found a relation between safety factors and the probability of bridge pier failure (see Table 2.2). However, whether these safety factors for the probability of failure are representative is not clear because the background of the factors is unknown.

Figure 2.5 shows the probability density function and the cumulative density function of the ratio between the measured and calculated equilibrium scour depth for about 750 plunge pool scour experiments (Hoffmans, 2012). About 50% computes a higher or lower value than the predicted one. If the accuracy of the predictor increases, the standard deviation, which measures here $\sigma = 0.25\ \mu$, decreases. Based on the experimental and computational results, the variation coefficient (ratio between the standard deviation and the mean value) is $\sigma/\mu = 0.25\mu/\mu = 0.25$. Hence, with this knowledge a safety factor can be determined; see also Section 2.5 where some examples are presented.

According to Breusers et al. (1977) the averaged scour depth at circular bridge piers is given as 1.4 times the pier diameter: $y_{m,e} = 1.4b$. Based on a systematic research on scour, Melville and Coleman (2000) found an upper limit of $y_{m,e,\max} = 2.4b$. Assuming that the experimental results are normally distributed, the difference between the upper limit (with a 99% confidence interval) and the mean value is about three times the standard deviation, thus $y_{m,e,\max} - y_{m,e}|_{50\%} = 3\sigma$ or $b = 3\sigma$ and thus $\sigma = 0.33b$ and thus the variation coefficient var $= \sigma/\mu = 0.33b/1.5b = 0.22$. Consequently, the variation coefficient for plunging jets and bridge piers is comparable.

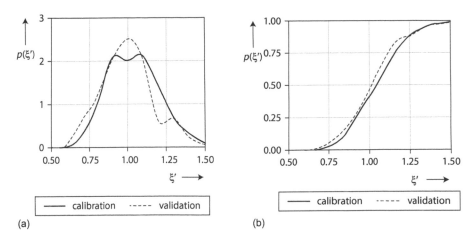

Figure 2.5 (a) $p(\xi')$ as a function of ξ' (b) $P(\xi')$ as a function of ξ' (Hoffmans 2012).

2.3.4 Failure probability approach

Methods to determine the failure probability are, for example, a Monte Carlo simulation, the FORM method, and numerical integration; see Melchers (2002) and Diermanse et al. (2015).

The reliability function Z (strength minus load) is the central point:

$$Z = R - S \tag{2.3}$$

The failure probability P_f can be determined if the reliability index β is computed:

$$\beta = \frac{\mu_Z}{\sigma_Z} \tag{2.4}$$

The Eurocode distinguishes three safety levels in which the Reliability Class (RC) and the Consequence Class (CC) are described for a reference period of 50 years. The following safety requirements are defined:

- RC1/CC1: $\beta = 3.3$ ($P_f = 4.83\text{E-}04$)
- RC2/CC2: $\beta = 3.8$ ($P_f = 7.23\text{E-}05$)
- RC3/CC3: $\beta = 4.3$ ($P_f = 8.54\text{E-}06$)

The ratio RC1/CC1 means that in case of a failure, the consequences are very limited; for RC2/CC2, the consequences are moderate; and for RC3/CC3, they are severe. More detailed descriptions for the CCs are as follows:

Table 2.4

Consequence Class	Consequences in terms of loss of human life	Economic, social or environmental consequences
CC1	Low	Small or negligible
CC2	Medium	Considerable
CC3	High	Very great

In general, hydraulic structures must be designed on the RC3/CC3 level.

Applying Equation (2.1) requires computing each component of scour, and this manual provides guidance. Using the probabilistic approach means that for all parameters, an average value and a standard deviation must be known or estimated beforehand (assuming that all parameters fulfil the standard normal distribution). Furthermore, mutual correlations between various parameters should be taken into account when relevant as this may influence the failure probability. In Section 2.5, examples are given.

2.4 Protective measures

2.4.1 Introduction

Several methods may be used to protect hydraulic structures from damage due to scouring. A conservative measure is to place the foundations of structures at such a depth that the deepest scour hole will not threaten the stability of the structure. Another way is to prevent the generation of erosive vortices. Hydraulic structures placed in waterways are often streamlined in order to reduce the drag exerted by flow and to reduce the effects of wake and turbulence intensity. Streamlining by means of deflectors and guide vanes, however, is effective only when the hydraulic structure is aligned with the flow within narrow limits.

From a geotechnical point of view, the stability of the upstream scour slope is of prime importance both during the time-dependent scour process and in the final situation when the equilibrium geometry has been attained. Besides sediment transport in the scour hole, soil particles in the filter structure below the bed protection can also be transported in both vertical and horizontal directions. If the groundwater seepage flow concentrates itself in narrow passages or pipes, the hydraulic structure may fail due to the transport of soil particles within the filter structure. Sand percolation hence merits special attention.

Furthermore, the stability of both the upper layer and the downstream termination of the bed protection against currents and eddies has to be safeguarded, so that the bed protection, and thus the hydraulic structure, will not be undermined.

2.4.2 Bed protection

Placing a bed protection downstream or around hydraulic structures is a common method of local scour protection. In principle, two types can be distinguished: the permeable, which is sand-tight, and the impermeable type. Furthermore, hard concrete bed protection close to the structure can be distinguished from the more flexible protections such as mattresses, sand bags and stones. Scour occurs in the area of the bed beyond the flexible bed protection, and as the scour hole is formed, the bed protection slides down into it. When rock mattresses or loose riprap are used, the possibility of erosion of fine particles from underneath the bed protection needs to be considered. This will be discussed in Section 2.4.3. Local scour can be reduced or prevented by either reducing the loading parameters or by increasing the strength parameters (Table 2.4).

In the 1990s, several highway bridges in the US collapsed due to insufficient (geotechnical) stability, following scouring at the bridge piers. After thorough investigations of the causes, possible countermeasures were also examined, e.g. a bed protection.

Table 2.4 **Protective measures**

Load reduction	Strengthening
Lengthening of bed protection	Compacting subsoil
Roughening of bed protection	Grouting subsoil
Streamlining of hydraulic structure (guide vanes, collars, deflectors)	Protecting bed or upstream scour slope (mattress protection, sand and stone bags, hinged concrete slabs, artificial seaweed, flexible mats)
Energy dissipators (berms, shallow foreland, vegetation, reed)	

A stable bed protection can be determined using appropriate design rules as provided by manuals, such as the Rock Manual (CIRIA/CUR/CETMEF, 2007) and 'The influence of turbulence on soil erosion' (Hoffmans, 2012). Two methods are available for determining the dimensions of the top layer:

1. Local average values of flow velocity and turbulence, reference level near the bed,
2. Depth-averaged values of flow velocity and turbulence.

For uniform flow, clear relations exist between local and depth-averaged flow velocity, and, consequently, both approaches can be applied. For non-uniform flow, approach 1 is physically more correct than approach 2 because local rather than depth-averaged conditions determine the load on the top layer material.

Design equations using local load parameters are the predictors of Jongeling et al. (2003, 2006), Hofland (2005) and Hoan (2008). In tender studies and pre-feasibility studies, local values will not be available in most situations because the determination of such values requires advanced models or experiments. Approach 1 is therefore less attractive for a first estimate of the dimensions of the top layer. In later stages, when optimising the top layer downstream of complex hydraulic structures, approach 1 is recommended. In these cases, additional research has to be conducted either numerically or experimentally.

Applying option 2, the Pilarczyk equation provided in the Rock Manual (CIRIA/CUR/CETMEF, 2007) and which is essentially the modified Isbash/Shields equation, or the Hoffmans equation can be used. Note that the Pilarczyk equation counts the turbulence twice: once with the so-called turbulence factor and once via the velocity factor. Because this is physically incorrect (see page 160 in Hoffmans 2012), the value of the turbulence is not always modelled correctly. The Hoffmans equation can be applied for both non-uniform and uniform flow (Hoffmans, 2010, 2012). Both equations are user-friendly.

Obviously, complex structures require carrying out laboratory tests or numerical computations as well as applying approach 1 with local values for flow velocity and turbulence in order to achieve a balanced design of the top layer (and filter layers).

All bank-hardening, embankment stabilising, and flow-altering countermeasures (except perhaps collars) listed in Table 2.4 can also be used for channel and bank stabilisation. A stable channel is paramount in protecting a bridge because if the channel migrates around and bypasses the bridge crossing, then a new bridge must be built.

2.4.3 Falling apron

A particular countermeasure is a falling apron. The function of a falling apron is to stabilise the end of a bed or slope protection to prevent undermining of the edge or to prevent sliding of the hydraulic structure into the scour hole. The principle is as follows (see Figure 2.6). At the edge of the protection extra gravel or rock will be installed (with mattresses extra length will be installed). The bed material will be exposed to the flow velocities, and a scour hole will start to develop. Gravel or rock of the protection rolls down over the slope due to gravity into the scour hole. When the scour hole reaches its equilibrium stage, the slope is expected to be protected sufficiently. With increasing scour depth, the scour hole slope will decrease.

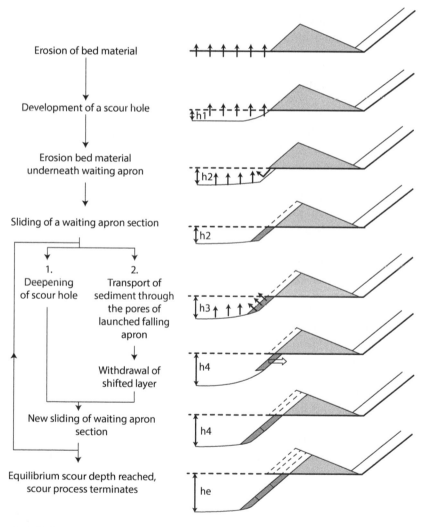

Figure 2.6 Behaviour of a falling apron (Delft Hydraulics, 2006).

The essential question is how much extra gravel or rock has to be installed in order to realise a minimal but stable protection. This requires insight into the scour process (scour depth and slope steepness), filter rules (sand is not allowed to be washed out due to internal erosion) and the failure mechanisms of slides.

Although the Rock Manual (CIRIA/CUR/CETMEF, 2007) presents a design methodology, physical tests showed that a single layer of rock will be formed after the falling apron has functioned. Such a single layer is insufficient to prevent washing out of fine particles. Thus, it was concluded that in most situations a falling apron will not function as expected. Regular monitoring and supplying extra rock will be necessary, implying that a falling apron always requires maintenance. How often maintenance is needed will depend on the frequency of its exposure to extreme hydraulic loads.

2.4.4 Other counter measures

A full range of available countermeasures is suitable for different situations. Proper classification is required to understand each one of them and to develop selection criteria. Countermeasures can be classified as follows:

- Flexible revetments or flexible bed armour: dumped rock riprap, rock-and-wire mattresses, gabions, planted vegetation, precast concrete blocks, willow mattresses.
- Rigid revetments or rigid bed armour: concrete pavement, sacked concrete, concrete-grouted riprap, concrete-filled fabric mats, bulkhead.
- Flow-control structures: groynes, retards, dikes, guide banks/bunds, check dams, jack fields, bendway weirs, hardpoints.
- Special devices: drift deflectors, abrasion armour at pier nose, bulkheads.
- Modifications of bridge, approach roadway, or channel underpinning or jacketing of pier, construction of outflow section on roadway, alignment of approach channel.
- Measures incorporated into the design of a replacement bridge: increased bridge length, fewer or no piers in the channel.

Countermeasures to arrest or retard meander migration include groynes, dikes, riprap, concrete pavements, bulkheads, guide banks and jack fields. Figure 2.7 shows the use of various types of containers for filling up scour holes in the river bed.

2.5 Examples

2.5.1 Introduction

Nowadays a risk assessment should play a role in the design of structures. In Section 2.5.2, the influence of the reliability index on the computation of the length of a bed protection downstream of a sluice with a reliability index is shown. In Section 2.5.3, the failure probability of a scour hole at a bridge pier is computed using the first-order mean value approach. In Section 2.5.4, the same scour hole is computed using a level II probabilistic method with a predetermined safety factor.

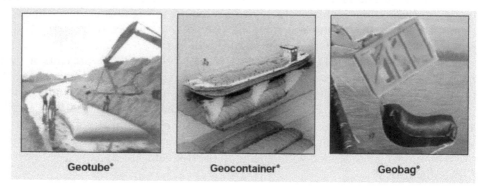

Figure 2.7 **Geotube, Geocontainer and Geobag (Deltares, 2015).**

2.5.2 Determination of the length of a bed protection with a reliability index

In this example, the bed protection downstream of a sluice is computed with a rule-of-thumb method and a probabilistic method, making the difference between the methods clear. The probabilistic approach results in a safety factor compared with the rule-of-thumb method.

The bed protection downstream of a sluice should be sufficiently long to prevent a shear failure:

$$L_{min} \geq L_{ins} \text{ or } Z = L_{min} - L_{ins}$$

Using the reliability Equation (2.3), the average value can be written as follows:

$$\mu_Z = \mu_R - \mu_S$$

or

$$\mu_Z = L_{min} - \mu\left(L_{ins}\right)$$

where
L_{min} = length of the bed protection (m)
L_{ins} = critical length (m)

The standard deviation of Z is defined as follows:

$$\sigma_Z = \sqrt{\left(\sigma^2_{L_{min}} + \sigma^2_{L_{ins}}\right)}$$

The length of the bed protection is considered as a deterministic parameter, i.e. $\sigma_{L_{min}} = 0$:

$$\sigma_Z = \sigma_{L_{ins}}$$

This results in the following reliability index:

$$\beta = \frac{\mu_Z}{\sigma_Z} = \frac{L_{min} - \mu(L_{ins})}{\sigma_{L_{ins}}} \qquad (2.4)$$

For shear failure, the mean actual bed protection length is (see Table 3.4):

$$L_{ins} = 1.7 y_{m,e|50\%}$$

where $y_{m,e|50\%}$ = equilibrium scour depth exceeded by 50% of the scour.

With $\sigma_{y_{m,e}} = \tfrac{1}{4} y_{m,e|50\%}$ (see also Section 2.3.3)

and thus $\sigma_{L_{ins}} = 1.7\sigma_{(y_{m,e})|50\%} = 1.7(\tfrac{1}{4} y_{m,e|50\%}) = 0.425 y_{m,e|50\%}$

this results into:

$$\beta = \frac{L_{min} - 1.7 y_{m,e50\%}}{0.425 y_{m,e50\%}} \quad \text{or} \quad L_{min} = (0.425\beta + 1.7) y_{m,e|50\%}$$

With a reliability index of $\beta = 4.3$, thus a failure probability of $P_f = 8.54 * 10^{-6}$ (see Section 2.3.4) the minimum bed protection length becomes

$$L_{min} = 3.53 y_{m,e50\%}$$

Conclusion: The bed protection length to prevent shear failure L_{ins} is 1.7 times the equilibrium scour depth based on a rule of thumb. However, using a probabilistic method with $\beta = 4.3$, the required length L_{min} is $3.53 y_{m,e|50\%}$. This implies a safety factor of $3.53/1.7 = 2.075$. The factor 2.075 proves to be a correct choice, but this is also because (1) $\sigma = \tfrac{1}{4}\mu$ and (2) a value of $\beta = 4.3$ is applied. Obviously, if the safety class is changed, the factor 2.075 will change too.

A probabilistic approach will be interesting if all components of an equation are considered in combination because in that case, mutual correlations will play a role.

2.5.3 Determination of the failure probability using a FORM approach

Assume that among all forms of scour in Equation (2.1), only the local scour, y_s, is considered, for example the scour at a bridge pier according to the HEC-18 equation (Equation 8.9):

$$y_{m,e} = 2.0 K h_0 F r^{0.43} \left(\frac{b}{h_0} \right)^{0.65} \qquad (8.9)$$

This equation gives an upper limit approach of the scour depth. In this example the coefficient 2.0 is considered as a safety factor making sure that the predicted scour will not be larger (note: the estimated value of 2 for the safety factor is perhaps too high, but this is not relevant for the goal of the example showing the FORM approach). This means that for computing the failure probability, the equation should be written as follows:

$$y_{m,e} = c K h_0 F r^{0.43} \left(\frac{b}{h_0} \right)^{0.65} \qquad (2.5)$$

With $\mu(c) = 1$ and $\sigma(c) = 0.333$ assuming that 100% of the scour depths are less than $\mu(c) + 3\sigma(c)$.

Substituting $Fr = \dfrac{U_0}{\sqrt{gh_0}}$ and $K = 1$, Equation (2.5) can be rewritten as:

$$y_s = cU_0^{0.43}h_0^{0.135}b^{0.65}g^{-0.215}$$

where
b = pier diameter
U_0 = average flow velocity
h_0 = average water depth
g = gravity acceleration
K = factor for various influences, such as pier shape, flow angle, etc
Fr = Froude number

Equation (2.5) is rewritten to translate this into the strength component R and the load component S of the reliability function Z. Afterwards, the average value and the standard deviation of all parameters are determined in order to compute the reliability index β and the failure probability P_f.

An alternative is not to divide the strength and load components but to compare the value of $y_{m,e}$ with a critical value $y_{m,e,cr}$. The reason for doing this is that some parameters represent the load and other parameters the strength. It is difficult to separate them, and the proposed alternative is a good option to deal with this problem. It is therefore often used.

The reliability function Z can be written with:

$$Z = R - S = y_{s,cr} - y_{s,actual} = y_{s,cr} - cU_0^{0.43}h_0^{0.135}b^{0.65}g^{-0.215} \tag{2.6}$$

Table 2.5 presents average values and standard deviation assuming normal distributions for all parameters.

Substitution of the average values gives an actual scour depth of $y_{s,actual} = 1.215\,\text{m}$.

In order to compute the reliability index β and the failure probability P_f, the values of μ_z and σ_z have to be computed. The reliability index β follows with Equation (2.4), whereas μ_z follows with Equation (2.6) by assuming a value for $y_{s,cr}$, for example $y_{s,cr} = 2.43\,\text{m}$ (the value that can be computed with $c = 2$). This results in $\mu_z = 2.43 - 1.215 = 1.215\,\text{m}$.

Table 2.5 Parameter values

Parameter	Average value	Standard deviation
c	1.0	0.333
b	1.25 m	0.10 m
U_0	2 m/s	0.25 m/s
h_0	6 m	0.25 m
g	9.8 m/s^2	0

The determination of σ_z is complicated and will be explained hereafter. The relevant equation reads $\sigma_z = \left\{ \sum_{i=1}^{n} \left(\dfrac{\partial Z}{\partial x_i} \sigma(x_i) \right)^2 \right\}^{0.5}$

In essence, it is a summation of the products of the standard deviations of each parameter and the derivative of Z in that parameter. This results for all parameters in

$$\frac{\partial Z}{\partial c} = U_0^{0.43} h_0^{0.135} b^{0.65} g^{-0.215}$$

$$\frac{\partial Z}{\partial U_0} = 0.43 U_0^{-0.57} c h_0^{0.135} b^{0.65} g^{-0.215}$$

$$\frac{\partial Z}{\partial h_0} = 0.135 h_0^{-0.865} c U_0^{0.43} b^{0.65} g^{-0.215}$$

$$\frac{\partial Z}{\partial b} = 0.65 b^{-0.35} c U_0^{0.43} h_0^{0.135} g^{-0.215}$$

$$\frac{\partial Z}{\partial g} = -0.215 g^{-1.215} c U_0^{0.43} h_0^{0.135} b^{0.65}$$

These formulas can also be written in another way, for example the equation for U_0:

$$\frac{\partial Z}{\partial U_0} = 0.43 U_0^{-0.57} c h_0^{0.135} b^{0.65} g^{-0.215} = \frac{0.43}{U_0} c U_0^{0.43} h_0^{0.135} b^{0.65} g^{-0.215} = \frac{0.43}{U_0} y_{s,act}$$

Substituting the relevant values, it is possible to compute the values of $\dfrac{\partial Z}{\partial x_i}$ for all parameters (see Table 2.6).

$$\sum_{i=1}^{n} \left(\frac{\partial Z}{\partial x_i} \sigma(x_i) \right)^2 = 0.172 \cdot 100\%$$

Now the value of σ_z reads $\sigma_z = (0.172)^{0.5} = 0.415\,\text{m}$.
With $\mu_z = 1.2145$, the reliability index β follows with Equation (2.4): $\beta = 2.93$.

Table 2.6 Characteristic values for the all parameters

X_i	$\mu(x_i)$	$\sigma(x_i)$	$\dfrac{\partial Z}{\partial x_i}$	$\left(\dfrac{\partial Z}{\partial x_i} \sigma(x_i) \right)^2$	Percentual contribution
c	1.0	0.333	1.2145	0.1636	95.1
B	1.25 m	0.10 m	0.6315	0.0040	2.4
U_0	2 m/s	0.25 m/s	0.2611	0.0043	2.5
h_0	6 m	0.25 m	0.0273	0.00005	0
g	9.8 m/s^2	0 m/s^2	0.0266	0	0

The failure probability P_f can be determined from a table with the standard normal or Gaussian distribution (see statistics manuals). This results for $\beta = 2.93$ in $P_f = 0.0017$ or 0.17%.

Suppose we did not apply the above method but a deterministic upper-limit approach with $c = \mu(c) + 3.\sigma(c) = 1.0 + 3 * 0.333 = 2$. Substituting the average values for the other parameters in Equation (2.5), we compute a scour depth of $y_s = 2.43$ m. The probabilistic approach shows that this value is related to 0.17% failure probability. Note that applying a $3.\sigma(c)$ does not really give an upper limit but a confidence of 99.5% because 0.5% of the values according to a standard normal distribution will exceed the value of the mean + three times the standard deviation. In other words, this deterministic approach leads to 0.5% failure probability.

In the above determination of the failure probability, we used average values. However, it is not recommended to determine the derivatives for the "average" situation, but for situations of potential failure of the system (in this example, situations where the critical value $y_{s,cr}$ will be exceeded). In general, the derivatives will therefore be determined for the "design point", which is the most likely combination of results of the parameters resulting in a failure. However, it is almost never possible to establish the design point analytically. Some iterative methods have been developed to localise the design point.

Table 2.6 shows that the standard deviation of the coefficient c gives the largest contribution to the failure probability, namely 95.1%. Increasing the accuracy of the parameter c with, for example small-scale laboratory research or analysis of observed scouring can improve the results.

The presented example is a simple method, according to the first-order second-moment probabilistic method, also called FORM, first-order mean-value approach (Melchers, 2002; Diermanse et al., 2015). It is one of the level II probabilistic calculation methods in which a linearisation is applied. In this example, the linearisation was performed around the average value. Other methods are a Monte Carlo approach or numerical integration. These are level III methods. A level II method results by definition in an approximation of the failure probability because mathematical descriptions of the failure mechanisms are linearised. With a level III method in principle, it is possible to determine a nearly exact estimate of the failure probability. For relatively simple models as described in the example, it is possible to linearise. The computation time with a level III method can be high for complex models.

2.5.4 Determination scour depth using a safety factor

Another way to calculate the expected scour depth would be to make use of a safety factor. The example compares the different results applying probabilistic methods with average values and characteristic values for various values of the safety factor. As calculated in the example above, the expected average scour depth is 1.215 m using the average values. With a critical scour depth of 2.43 m, this calculation resulted in $P_f = 0.0017$ or 0.17%. If the average scour depth ($y_{s,actual}$) is considered invariable, the critical scour depth can be used as a parameter.

If the failure probability should be equal to 10%, the critical scour depth can be calculated using Equation (2.3). A reliability of 10% is equal to a β value of 1.28, assuming a normal distribution. With $\sigma = 0.415\ \mu$, see Section 2.5.3, follows with $\beta = \dfrac{\mu_z}{\sigma_z}$ thus $\mu_z = \beta \cdot \sigma_z = 1.28 \cdot 0.415 = 0.531$ m. Hence, the critical scour depth should

Table 2.7 Scour depth with certainty when the average scour depth is 1.215 m

Safety factor Johnson	Failure probability	β	μ (m)	Critical scour depth using probability method (m)	Critical scour depth using safety factor Johnson factor with average value (m)	Critical scour depth using safety factor Johnson factor with characteristic value (m)
1.0	0.5	0	0	1.21	1.21	1.90
1.2	0.1	1.28	0.53	1.75	1.46	2.28
1.4	0.01	2.33	0.97	2.18	1.70	2.66
1.6	0.001	3.09	1.28	2.50	1.94	3.04
1.75	0.0001	3.71	1.54	2.75	2.13	3.32
1.80	0.00001	4.3	1.78	3.00	2.19	3.42

be $1.215 + 0.531 = 1.746$ m. If the same calculation is carried out with a safety factor as proposed by Johnson (1992) using Equation (2.2), the safety factor of a failure probability of 10% is equal to 1.2 (see Table 2.2). Accordingly, if the average scour depth is equal to 1.215 m the design scour depth would be $1.215 \times 1.2 = 1.458$ m. However, this calculation was carried out with the average value, a better (more realistic to normal calculations) comparison is to calculate this with the characteristic value, which would result in $(1.215 + 1.65 \times 0.415) \times 1.2 = 2.280$ m. These calculations were made for multiple probability failure rates, as shown in Table 2.7. The values used for the parameters in Equation (2.6) are shown in Table 2.5. When these values are combined with the safety factor of Johnson, a characteristic value of $1.25 \times 1.65 = 2.06$ m and beta from a normal distribution, see Table 2.7, can be calculated.

As shown in this example, the probabilistic method can prove to be a useful tool to remove some conservatism.

Chapter 3

Design tools

3.1 Introduction

The stability of structures can be endangered due to scour or geotechnical processes. Scouring occurs when the local flow velocity or turbulence characteristics exceed the critical values. Under these conditions, bed material is eroded. Presence of a structure a decreasing dimension of the flow channel results in higher flow velocity and lower turbulence intensities. These aspects are addressed in Sections 3.2–3.4. Section 3.2 discusses the possibilities provided by using the various available mathematical scour and erosion models. Sections 3.3 and 3.4 discribe general and local scours, respectively.

Geotechnical instability is the result of various phenomena such as flow slides and shear slides and groundwater flow. The slides create sudden movements of large amounts of soil. Important are the slopes of scour holes and embankments. Geotechnical aspects relevant for the stability of structures are described in Section 3.5.

Hydraulic structures that obstruct the flow pattern in their vicinity may cause localised erosion or scour. Changes in flow characteristics (velocities or turbulence) lead to changes in sediment transport capacity and hence to a local disequilibrium between incoming and outgoing sediment transport. A new equilibrium may eventually be reached as hydraulic conditions are adjusted through scour. Scour at structures can be divided into general and local scours. These possible processes have different lengths and time scales. As the first approximation, the scour caused by each process separately may be added linearly to obtain the resulting scour. In addition, scour in different conditions of sediment transport can be distinguished. In general, the time scale of local scour is relatively short for example for bridge piers. However, at sills in prototype situations, the equilibrium scour depth can be reached after several years and for very large flow depths, say 50 m, after decades. These morphological aspects are considered in this chapter.

3.2 Mathematical scour and erosion models

3.2.1 Introduction

In the early 1990s, a morphological model of the development of scour holes behind hydraulic structures was developed (Hoffmans, 1992). This morphological model, which is based on the Navier–Stokes and convection–diffusion equations, can simulate two-dimensional scour downstream of a sill with a bed protection. In the scour hole, different zones were distinguished in which the turbulence was prescribed with analytical relations obtained from the literature.

At present (2020), several models are available to predict two- and three-dimensional scour. The first more or less traditional category of these models consists of two separated sub-models: a flow model and a morphological model which are connected to one model. The flow model calculates the flow velocities, the morphological model predicts the sediment transport (both suspended and bed load) and the bed level changes as a function of time.

Although these models predict scour parameters, they do not always predict the scouring process correctly due to the poor modelling of the turbulence in the scour hole and the interaction between water and soil that governs the erosion processes. For example, in shallow water modelling, the turbulence structures are not resolved, and the velocity is integrated along the water column. Hence, it is expected that shallow water modelling is unable to simulate such a process.

Reynolds-Averaged Navier–Stokes (RANS) models offer the most economic approach for computing complex turbulent flows. Typical examples of such models are the k-epsilon or the k-omega models in their different forms. These models simplify the problem to the solution of two additional transport equations and introduce an eddy-viscosity (turbulent viscosity) to compute the Reynolds Stresses.

Direct Numerical Simulation (DNS) and Large-Eddy Simulation (LES) are alternative approaches for reproducing the turbulence in a better way. It is expected that DNS yields best results. Unfortunately, DNS is computationally demanding and is limited to small spatial scales and low Reynolds numbers on nowadays computational resources (Kidanemariam & Uhlmann, 2014). LES is computationally less demanding and can be applied for high Reynolds numbers. Hence, LES can be a suitable approach for simulation of the scouring process (Nabi et al., 2013b).

A second category of these models consists of particle models or multiphase models, where flow and sediments are solved together as different phases. These are mesh-free approaches, where flows and sediments are calculated as particles. An example of such an approach is a two-phase numerical model using Smoothed Particle Hydrodynamics (SPH) (e.g. Gingold & Monaghan, 1977; Monaghan, 1992), which is applied for scour frequently. Another example with possible application to scour modelling is the Material Point Method (MPM) which has been developed for geotechnical applications. Also, MPM is applicable for modelling complex soil-water interactions that determine scour.

3.2.2 *Types of modelling*

Three types of mathematical models can be distinguished; see also Figures 3.1 and 3.2 where the differences between these models are shown:

1. Large-scale RANS models: shallow water and turbulence modelling;
2. High-resolution hydrodynamic models: LES and DNS;
3. Particle-based multiphase models: Hydraulic model: SPH; Soil mechanic model: MPM.

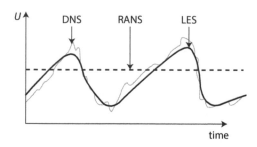

Figure 3.1 **Principle of the RANS, LES and DNS approaches. Flow velocity in the streamwise direction as a function of time for three different models. (Adapted from Nicoud, 2007.)**

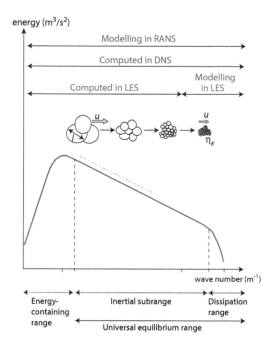

Figure 3.2 **Turbulent kinetic energy spectrum versus wave number. The level of computed and modelled scales in RANS, DNS and LES. (Adapted from Sagaut et al., 2006.)**

3.2.3 *Large-scale RANS models*

3.2.3.1 *Shallow water modelling*

Some examples of RANS models are the mathematical modelling systems Delft3D (Deltares), Mike software of DHI, Finel of Svasek Hydraulics, Tuflow and Telemac–Mascaret. These systems consist of different models. One sub-model solves the unsteady shallow water equations in two (depth-averaged) or in three dimensions. The system of equations consists of the continuity equation and the two or three equations of motion. The equations are formulated in orthogonal curvilinear co-ordinates in which the free surface level and bathymetry are related to a flat horizontal plane of reference.

For gradual transitions and/or streamlined structures, these sub-models calculate the flow velocities satisfactorily. However, in non-uniform flow zones, e.g. downstream of backward facing steps where the flow separates the velocities are predicted insufficient. Since turbulence parameters lack in these models, the scouring or erosion process as a function of time is not well predicted and therefore they are only useful for the computation of boundary conditions of scour equations.

3.2.3.2 Turbulence modelling

Typically, the flow in a scour hole is computed by using a turbulence model. The most well-known turbulence models for the RANS equations are the k-epsilon and k-omega models which simulate the mean flow characteristics for turbulent flow conditions. These models, which have also been implemented in Delft3D and the aforementioned models, give a general description of turbulence by means of two transport equations. Jacobsen et al. (2014) present scour analysis at a pile by using the k-omega model.

In RANS models, the turbulence is smoothed out by an averaging operation, and hence this class of modelling is unable to produce the bed scouring correctly (Khosronejad et al., 2012). Solving this process, more sophisticated modelling is necessary at which the turbulence structures can be resolved accurately.

3.2.4 High-resolution hydrodynamic models

3.2.4.1 Hydrodynamic model LES

LES is a mathematical model for turbulence used in computational fluid mechanics and is currently applied in a wide variety of engineering applications. The simulation of turbulent flows by numerically solving the Navier–Stokes equations requires resolving a very wide range of time and length scales, all of which affect the flow field.

The principal idea behind LES is to reduce the computational cost by ignoring the smallest length scales (see also Figure 3.2), which are the most computationally expensive to resolve, via low-pass filtering of the Navier–Stokes equations. Such a low-pass filtering, which can be viewed as a time- and spatial-averaging, effectively removes small-scale information from the numerical solution.

LES directly calculates the motion of the largest turbulent eddies in at least a portion of the domain and therefore suitable for predicting both the flow velocities and the turbulence parameters and thus sediment entrainment and sediment transport in non-uniform flow and thus scouring/erosion as a function of time.

3.2.4.2 Application of LES

Nabi et al. (2012) introduced a model based on LES for flow modelling. The sediment transport is then resolved by considering the sediment as spherical rigid particles, and the particles are traced in a Lagrangian framework (Nabi et al., 2013a). The bed morphology is then calculated based on particles (sediment) erosion and deposition. This model showed its great success in simulating the small-scale bed morphodynamics such as dunes and pier scouring (Nabi et al., 2013b). The model was later extended by Kim et al. (2014) in simulating the scouring generated by two adjacent piers. Figure 3.3 shows the flow and the scouring in its equilibrium state.

This work was further extended by Kim et al. (2015) for simulation of patch of piers (model for vegetation). Figure 3.3 also shows the bed elevation under turbulent flow passing throw a group of piers. This model showed success in simulation and studied the effect of pier interactions on bed pier scouring.

3.2.4.3 Hydrodynamic model DNS

DNS is a simulation in computational fluid dynamics in which the Navier–Stokes equations are numerically solved without any turbulence model. This means that the whole range of spatial and temporal scales of the turbulence must be resolved. All the spatial scales of the turbulence must be resolved in the computational mesh, from

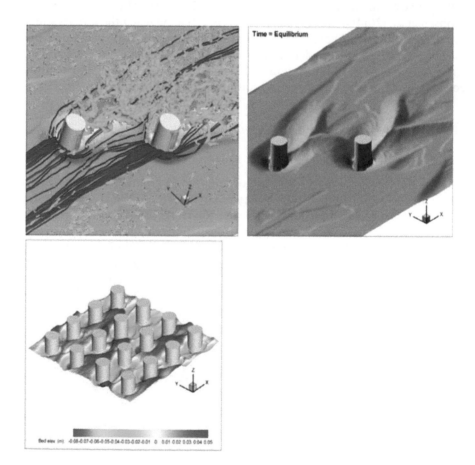

Figure 3.3 Computational results of Nabi's modelling.

the smallest dissipative scales (Kolmogorov microscales), up to the integral scale, associated with the motions containing most of the kinetic energy (see also Figure 3.2).

The computational cost of DNS is very high, even at low Reynolds numbers. For the Reynolds numbers encountered in most industrial applications, the computational resources required by a DNS would exceed the capacity of the most powerful computers currently available. However, direct numerical simulation is a useful tool in fundamental research in turbulence. Using DNS, it is possible to perform "numerical experiments" and extract from them information difficult or impossible to obtain in the laboratory, allowing a better understanding of the physics of turbulence.

Summarising DNS numerically solves the full unsteady Navier–Stokes equations and calculates the exact turbulence solution. It is a useful model; however, it is too computationally intensive to be practical.

3.2.5 Particle-based multiphase models

3.2.5.1 Soil mechanics: MPM

MPM is a numerical technique used to simulate the behaviour of soil, water and air. In the MPM, a continuum body is described by a number of small Lagrangian elements referred to as "material points". These material points are surrounded by a background mesh/grid that is used only to calculate gradient terms such as the deformation gradient.

Large deformation and soil–water–structure interactions exist in many environmental and civil engineering problems, such as landslides, flow slides and shear failures, installation of piles in saturated soils, settlement due to consolidation processes, fluidisation and sedimentation processes in submerged slopes and internal erosion in dykes. Modelling these processes is challenging due to hydro-mechanical coupling, large deformation and contact problems.

Recently, Yerroa et al. (2017) presented a first approach of the MPM to model the internal erosion process in internally unstable soils. These soils consist of a mixture of coarse and fine particles, in which the coarse fraction forms the stable skeleton of the soil, whilst fine particles can be eroded and are able to move freely as a result of seepage flow. The computational results of the vertical groundwater flow are discussed. The calculations show that the fine fraction is eroded from the skeleton depending on the erosion law, and it transported as the fluidised material through the saturated porous media.

Currently, only laminar flow of free surface water and internal erosion in unstable soils are considered with the MPM. The description of scour under turbulent flow has to be done in future research programmes. Therefore, turbulence models (k-epsilon and k-omega models) and or LES models are still needed for predicting scouring near hydraulic structures. More information can be found in the study by Zhang et al. (2016).

3.2.5.2 Hydraulic model: SPH

Following de Wit (2006), SPH can be used in many different situations in hydraulic engineering. Especially in problems with large pressure gradients, fast varying water levels and intersecting free surfaces, the advantages of SPH show to full extent. Unfortunately a small time step is needed for stability, together with many particles needed for enough resolution; this leads to considerable calculation times. The application of SPH is therefore restricted to local and short phenomena. SPH can be used in situations where many other methods fail, for instance wave overtopping can be simulated in great detail. Although SPH is a relatively new technique in the field of hydraulic engineering, its future looks promising.

3.3 General scour

3.3.1 Introduction

The time scale to reach equilibrium for general scour is usually longer than the time scale for local scour. Commonly occurring examples of general scour are the long-term change in the bed level of a river, scour due to a long constriction, scour in a bend or scour at a confluence.

3.3.2 Overall degradation or aggradation

When conditions to which a river has been adjusted change, overall degradation or aggradation occurs. Examples of changes in conditions are increasing flow or increasing sediment transport in a river. Some examples of features, resulting either from human interference or from natural changes, to which the flow regime could respond, are as follows:

- Flood embankments, flood detention basins and weirs;
- Channel improvement schemes involving dredging or weed clearance;
- Mining of sand and gravel;
- Changes in water patterns (confluence or bifurcation of river channels);
- Schemes for transfer of water between river basins;
- Meander cut-offs.

Overall degradation at, for example, a bridge site lowers the bed level and may thereby increase the risk of failure of the foundations. Aggradation, on the contrary, will cause higher water levels and may reduce the risk from scour. Degradation processes already in progress have to be considered, as well as the possibility of inducing degradation processes in the future (e.g. seasonal degradation). Overall degradation can affect a long reach of the river, extending over tens to hundreds of kilometres and over periods of decades to centuries (de Vries, 1975). Examples are the Bovenrijn and Waal branches of the river Rhine in the Netherlands which experience general bed degradation due to enhanced sediment transport capacity and reduced sediment supply from upstream.

Information on potential general river bed changes can be obtained from a one-dimensional numerical model of the river morphology. Besides numerical models, analytical models can also provide insights into the nature of morphological processes. De Vries (1993) gives a useful introduction to the modelling of morphological changes in rivers due to natural causes or human interferences.

3.3.3 Constriction scour

Constriction scour (Figure 3.4) occurs in confined sections of a river and results in a lowering of the bed level across the width of the river. The increase in depth over a long constriction can be easily computed from the equations of motion and continuity for sediment and water. For the condition of general movement, the solution of Straub in 1939 reduces to (Laursen & Toch, 1956):

$$\frac{y_{m,e} + h_0}{h_0} = \frac{1}{(1-m)^\beta} \tag{3.1}$$

in which

$y_{m,e}$ = equilibrium scour depth in constriction (m)
h_0 = flow depth upstream of the constriction (m)
$m = 1 - B_2/B_1$, constriction ratio (–)
B_1 = width of the river upstream of the constriction (m)
B_2 = width of the river at the constriction (m)
β = exponent, which lies between 0.67 and 0.8

If there is bank overflow with discharge Q_f, Equation (3.1) becomes (Laursen & Toch, 1956):

$$\frac{y_{m,e} + h_0}{h_0} = \frac{1}{(1-m)^\beta} \left(\frac{Q}{Q - Q_f} \right) \quad \text{with} \quad Q = Q_c + Q_f \tag{3.2}$$

in which

Q = total discharge [m³/s]
Q_f = floodplain discharge [m³/s]
Q_c – discharge through main river (without floodplain); see Figure 3.4 [m³/s].

Equation (3.1) is generally applicable for constriction scour and based on the continuity equation and the Chézy equation. Equation (3.2) is an extended form of (3.1) and also generally applicable.

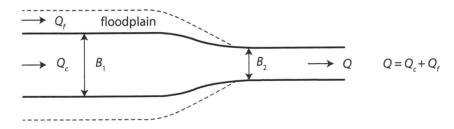

Figure 3.4 Constriction scour.

3.3.4 Bend scour

Scour in the outer parts of bends depends on local parameters (e.g. bend curvature, flow depth variation and grain size distribution) and upstream influences (redistribution of flow and sediment transport). This bend scour occurs as the result of spiral flow. Struiksma et al. (1985) showed that an overshoot phenomenon adds deeper scour to the scour resulting from spiral flow. The topography of the river bed adjusts to changing conditions by a damped response, overshooting the fully developed solution (Figure 3.5). The magnitude of this overshoot depends on the width-to-depth ratio: the overshoot increases with increasing width-to-depth ratio. Due to this overshoot effect, it is difficult to formulate a simple predictor for bend scour, although such a predictor is often needed to provide a first estimate. In such cases, local parameters have to be used, which implies that the fully developed bend solution (Odgaard, 1981) has been adapted. An example of this method is shown in Figure 3.5.

The time scale $T_{o,s}$ for the changes of the cross-sectional profile can be given by:

$$T_{o,s} = 0.85 \frac{B^2 \sqrt{\Psi}}{\pi^2 q_s} \tag{3.3}$$

in which

q_s = sediment transport per unit width (m^2/s)
Ψ = Shields parameter (–)
B = width (m)

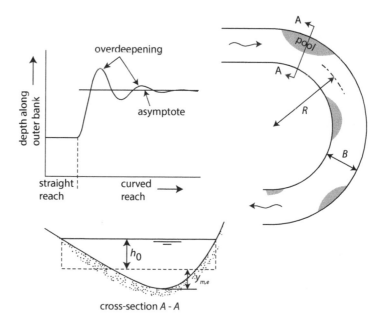

cross-section A - A

Figure 3.5 **Overdeepening of bathymetry in a river bend.**

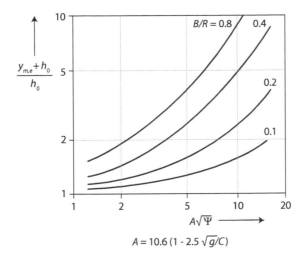

$$A = 10.6 \, (1 - 2.5 \sqrt{g}/C)$$

Figure 3.6 **Bend scour (Struiksma & Verheij, 1995).**

Equation (3.3) is generally applicable for bend scour conditions and based on the spiral flow in river bends. The formula has been validated by Struiksma and Verheij (1995), see Figure 3.6.

Following Thorne (1993), the equilibrium scour depth at a bend can be given by the empirical relation:

$$\frac{y_{m,e}}{h_0} = 1.07 - 0.19 \cdot \log\left(\frac{R}{B} - 2\right) \quad for \quad 2 < \frac{R}{B} \leq 22 \tag{3.4}$$

in which R is the radius of curvature of the centre-line.

Equation (3.4) is based on both flume experiments and prototype experiments in large rivers (flow depths up to 17 m), in which the mean particle diameter varied from 0.3 to 63 mm. The error between the measured scour and the scour predicted by using the empirical relation was found to be on average 25%. For a first estimate, the equilibrium scour depth in the bend can be assumed to be equal to the flow depth, which is a somewhat conservative estimate if values of R/B are large.

At the toe of river dikes, local scour occurs in outer river bends. The bend scour can undermine the dike in an outer bend; see the left sketch of Figure 3.7. In the natural situation, the main channel of the river moves towards the outer bend, forming meander bends. In the Netherlands, the rivers have been trained by using groynes, which limit the outward movement of the main channel.

In the Dutch situation and in the situation with a protected dike, locally more extreme bend scour can occur in the outer bend due to for example the extra turbulence induced by the bank protection or the impingement of a channel (Mosselman et al., 2000). For making a bank protection design, the excess bend scour at protected banks has to be taken into account, for example, by constructing a bottom protection, a falling apron (see Section 2.4), or another adaptive strategy.

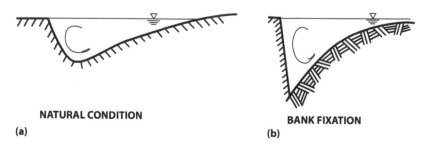

NATURAL CONDITION   **BANK FIXATION**

(a)   (b)

Figure 3.7 **Bend scour in a natural condition and with a fixed bank.**

More information can be found in Jansen et al. (1979) and in Hydraulic Engineering Circular No. 23 (HEC-23, 2009).

3.3.5 Confluence scour

When two branches of a river meet (see Figure 3.8), the water level gradient upstream of the confluence, the flow velocity direction and the flow velocity magnitude may differ. The mixing of the two water bodies results in enhancement of turbulence, which generates scour at the meeting point: confluence scour. Though mathematical models are available at present, for reasons of simplicity, the scour downstream of a confluence is related to the following variables (e.g. Breusers & Raudkivi, 1991):

$$\frac{y_{m,e}}{h_0} = c_0 + 0.037\theta \qquad (3.5)$$

in which
c_0 = coefficient depending on material properties (–), c_0=0.29
θ = angle between the two upstream branches (°)

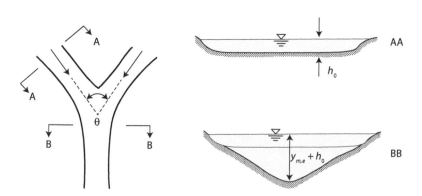

Figure 3.8 **Water depth at confluences.**

The value of the coefficient c_0 is 0.29 for the Jamuna river in Bangladesh with fine sand (Klaassen & Vermeer, 1988). Based on field data for gravel rivers, Ashmore and Parker (1983) found that c_0 is about 1.24. Ashmore and Parker (1983) presented similar results.

The empirical Equation (3.5) is based on a limited number of observations in the Jamuna River, and, consequently, validated in a limited way.

3.4 Local scour

3.4.1 Introduction

Local scour results directly from the impact of the structure on the flow. Physical model testing and prototype experience have permitted the development of methods for predicting and preventing scour at different types of structures. Data such as scour depth and location of deepest point with respect to scour can be obtained by testing physical models, and this approach may be particularly appropriate for unusual structures not covered by existing formulae or for field measurements of scour at existing structures.

Physical models are used in combination with computational results of horizontal (two-dimensional depth-averaged) numerical models to predict scour at locations with complex (three-dimensional) flow. The accuracy of scour computation mainly depends on the accuracy of the measurements of flow velocities and turbulence intensities. According to Breusers (1966), the development of the scour process depends on the flow velocity and turbulence intensity at the transition between the fixed and the erodible bed. By applying this concept, the scour prediction can be restricted to a single computation. No information is needed concerning the near-bed velocities and bed turbulence in the scour hole.

In principle, only the maximum scour depth in the equilibrium phase is relevant when dealing with scour problems. This is especially true for isolated structures such as bridge piers, spur dikes, groynes and other permanent structures (sills, weirs, final closure works). However, there are cases in which the time factor is important, for example, in the case of closure of estuary branches.

Based on clear-water scour experiments, i.e. no upstream sediment transport, using scale models with small Froude numbers (Breusers, 1966; Dietz, 1969), Zanke (1978) distinguished four phases in the evolution of a scour hole (Figure 3.9): an initial phase, a development phase, a stabilisation phase and an equilibrium phase.

In the initial phase, the flow in the scour hole is nearly uniform in the longitudinal direction. This phase of the scour process can be characterised as the phase in which the erosion capacity is most severe. Observations with fine sediments (e.g. Breusers, 1966) showed that at the beginning of the scour development, some bed material near the upstream scour slope goes into suspension. Most of the suspended particles follow convectional paths within the main flow and remain in suspension due to the internal balance between the upward diffusive flux and the downward flux due to gravity. Some of the particles will settle and will be resuspended due to the large bursts of the turbulent flow near the bed. Other particles that jump less high than a defined saltation or reference height are transported as bed load.

During the development phase, the scour depth increases considerably, but the shape of the scour hole does not change much. In this phase, the ratio of the maximum scour

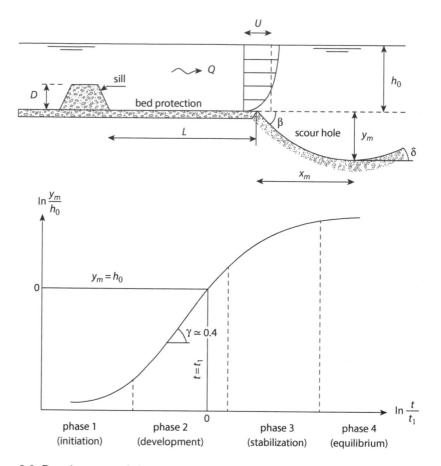

Figure 3.9 Development of the scour process.

depth to the distance between the end of the bed protection and the point of maximum scour depth is more or less constant. Measurements by Hoffmans (1990) showed that the upper part of the upstream scour slope is in equilibrium, whereas the lower part is still developing. The suspended load close to the bed decreases significantly compared to the condition in the initial phase. This can mainly be ascribed to the decrease in the flow velocities near the bed over time, despite the increase of the turbulent energy. Though bed particles are picked up and carried by the flow, the time-averaged value of the sediment transport at the upper part of the upstream scour slope is negligible since instantaneous velocities in upstream and downstream direction produce approximately equal contributions to sediment transport.

In the stabilisation phase, the rate of development of the maximum scour depth decreases. The erosion capacity in the deepest part of the scour hole is small compared to the erosion capacity downstream of the point of reattachment, so that the dimensions of the scour hole increase more in the longitudinal direction than in the vertical direction. The more the scour process continues, the more the flow velocities above the lower part of the upstream scour slope decrease. In the stabilisation phase, the equilibrium situation for both the upstream scour slope and the maximum scour depth is almost achieved. The equilibrium phase can be defined as the phase in which the dimensions of the scour hole do no longer change significantly.

From model tests on different scales and with different bed materials, relations were derived between the time scale and the scales for velocity, flow depth, and material density (Breusers, 1966, 1967; Dietz, 1969; van der Meulen & Vinjé, 1975). The main conclusions were that the shape of the scour hole is independent of the bed material and flow velocity.

3.4.2 Time-dependent scour

The scour process as a function of time can be given with reasonable accuracy, provided the equilibrium scour depth can be predicted satisfactorily:

$$\frac{y_m}{y_{m,e}} = 1 - e^{-\frac{\lambda}{y_{m,e}}\left(\frac{t}{t_1}\right)^{\gamma}} \tag{3.6}$$

in which
  t = time (s)
  t_1 = characteristic time at which $y_m = \lambda$ (s)
  y_m = maximum scour depth at t (m)
  γ = coefficient (–), γ = 0.4–0.8
  λ = characteristic length scale (m)

In the development phase (i.e. when $t < t_1$), Equation (3.6) can be approximated into:

$$\frac{y_m}{y_{m,e}} = 1 - \left(1 - \frac{\lambda}{y_{m,e}}\left(\frac{t}{t_1}\right)^{\gamma}\right)$$

or

$$\frac{y_m}{\lambda} = \left(\frac{t}{t_1}\right)^{\gamma} \tag{3.7}$$

which is equal to the original time-dependent scour formula of Breusers. With Equation (3.6), the maximum scour depth can be predicted for all phases in the scour process (development phase, stabilisation phase and equilibrium phase).

Breusers validated Equation (3.7) extensively in a systematic research project. The laboratory experiments showed that no scale effects occurred. This indicates that the same equation can be applied in prototype conditions.

Values of γ range from 0.2 to 0.4 for two-dimensional flow and up to 0.8 for three-dimensional flow. More complicated functions were required to describe the three-dimensional case.

The former Scour Manual (Hoffmans & Verheij, 1997) presented the following equation for the scour process as a function of time:

$$\frac{y_m}{y_{m,e}} = 1 - e^{\ln\left(1 - \frac{\lambda}{y_{m,e}}\right)\left(\frac{t}{t_1}\right)^{\gamma}}$$

(3.8)

Equations (3.6)–(3.8) are all shown in Figure 3.10 for an example with a scour depth of 2 m at $t = t_1 = 5$ h. The equilibrium scour depth $y_{m,e} = 2.5$ m. Though this example shows no large differences between Equations (3.6) (red line) and (3.8) (green line) Equation (3.8) gives no realistic results for equilibrium scour depths $y_{m,e}$ smaller than the scour depth y_m at $t = t_1$. Therefore, Equation (3.8) is replaced by Equation (3.6) (red line).

The comparison in Figure 3.10 shows the importance of knowing the correct length scale λ. Many definitions of characteristic length scales can be found in the literature, including, for example, the Kolmogorov length scale which represents the micro-turbulent eddies. The size of these eddies is in the range of 0.1–1 mm. The largest eddies play an important part in any turbulent flow. In shallow water conditions, the size of the largest (macro-turbulent) eddy is on the order of h_0, which can be used as the characteristic length scale λ. For deep water conditions, the dimensions of the largest eddies are nearly equal to those of the hydraulic structure (e.g. bridge pier width, $\lambda = b$, Chapter 8).

The time scale according to Breusers can be represented by (λ = flow depth):

$$t_1 = \frac{K h_0^2 \Delta^{1.7}}{(\alpha U_0 - U_c)^{4.3}} \quad \text{and} \quad \alpha = 1.5 + 5 r_0$$

(3.9)

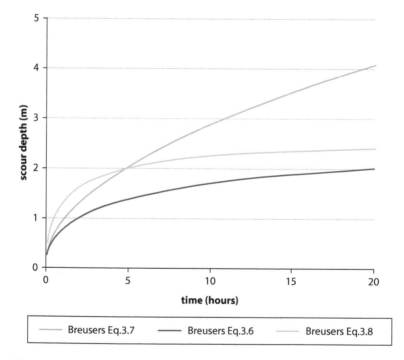

Figure 3.10 Comparison between various Breusers equations.

where

$K = (330)$ non-dimensionless constant $(\text{m}^{2.3}/\text{s}^{3.3})$
r_0 = depth-averaged relative turbulence-intensity upstream of the scour hole (-)
U_0 = depth-averaged flow upstream of the scour hole (m/s)
U_c = critical depth-averaged flow velocity (m/s)
α = turbulence coefficient (-)
Δ = relative density of bed material (-)
t_1 = characteristic time (hours)

So, t_1 is proportional to:

$$t_1 \propto h_0^2 \left[\alpha U_0 - U_c \right]^{-4.3}$$

For scale modelling, another way of writing the time scale reads:

$$n_t = n_\lambda^2 n_\Delta^{1.7} \left(n_{\alpha U_0 - U_c} \right)^{4.3} \tag{3.10}$$

in which n = scale ratio (parameter value in the prototype divided by parameter value in the model).

Equation (3.10) represents the time scale of scour. It essentially shows the relation between the scour hole volume and the sediment transport. However, the sediment transport description does not fulfil the Partheniades erosion law, and therefore Equation (3.10) should not be applied. A better equation for the time scale, although not calibrated and validated, reads (Hoffmans, 2012):

$$t_1 \propto h_0^2 \left[\frac{\vartheta U_0^2 - U_c^2}{U_c^2} \right]^{-2} \tag{3.11}$$

The parameter ϑ represents the influence of both the turbulence and the heterogeneity of the subsoil. It is strongly recommended to develop this equation using the systematic research at Delft Hydraulics in the period from 1960 to 1980.

3.4.3 Equilibrium scour

"This manual highlights not only the time-dependent Breusers method but also the equilibrium scour depth, which is important for safe designs. Hoffmans (2012) derived new formulas based on an analysis of equilibrium scour depths and the erosion law of Partheniades. The equilibrium scour depth for clear-water scour is:"

The equilibrium scour depth for clear-water scour is:

$$\frac{y_{m,e}}{\lambda} = A\chi_e \left(\frac{U_0}{U_c} \right)^2 - B \quad \text{for} \quad U_0 < U_c \tag{3.12}$$

And for live-bed scour,

$$\frac{y_{m,e}}{\lambda} = A\chi_e - B \quad \text{for} \quad U_0 \geq U_c \tag{3.13}$$

Table 3.1 Length scales and values for parameters A and B per type of construction (h_0 = initial water depth (m), b = width of bridge pier (m))

	λ	A	B
Sills	h_0	1.0	1.0
Bridge pier (slender)	b	1.6	1.3
Abutments, Bridge pier (wide)	h_0	1.4	1.0

in which:

$$\chi_e = \frac{1+6.3r_0^2}{1-6.3r_{0,m}^2} \tag{3.14}$$

with values for the parameters A and B given in Table 3.1,
  where

λ = length scale (m); see Table 3.1
r_0 = depth-averaged relative turbulence intensity at water depth change (-)
$r_{0,m}$ = depth-averaged relative turbulence intensity when scour depth is maximal (-)
U_0 = depth-averaged flow velocity above the bed (m/s)
U_c = critical depth-averaged flow velocity above the bed (m/s)
χ_e = turbulence parameter (-)

In the deepest part of the scour hole, $r_{0,m}$ can be approximated by:

$$r_{0,m} = \sqrt{\tfrac{1}{2}C_k}\left(\frac{y_m}{h_0}+1\right) \quad \text{for} \quad 0.1 < \frac{y_m}{h_0} < 2 \tag{3.15}$$

in which
  y_m = maximum scour depth (m)
  C_k = constant dependent on the steepness of the upstream slope (-), $0.03 - 0.045$

As a first estimate for the equilibrium scour depth, a default value of 0.25 can be used for $r0_{,m}$.
  Dietz (1969) applied the continuity equation to describe the scour hole. He derived the equilibrium scour depth:

$$\frac{y_{m,e}}{h_0} = \frac{\omega U_0 - U_c}{U_c} \tag{3.16}$$

Q = discharge (m^3/s)
A = cross section (m^2)
With $\omega = 1 + 3r_0$, r_0 is relative turbulence intensity (–)

Measurements by Dietz (1969) showed that the average value of the turbulence coefficient ω was about $\omega = 2/3 + 2r_0$. Dietz reported that the maximum value of ω amounted to $\omega_m = 1 + 3r_0$. Following Popova (1981; in Hoffmans & Verheij, 1997), the turbulence coefficient is given by $\omega = 0.87 + 3.25 \ Fr + 0.3r_0$ (Fr = Froude number just before the scour hole). According to Rossinskiy (1956; e.g. Blazejewski, 1991), ω lies in the range of 1.05–1.7, depending on the geometry upstream of the scour hole. Upstream sediment supply or the presence of more resistant layers will reduce the maximum scour depth. If sediment is supplied from upstream, values predicted by Equation (3.16) seem to be too high in the prototype situation.

The rate at which the scour hole approaches an equilibrium phase depends strongly on the turbulence intensity in non-uniform flow and, to a lesser extent, on material characteristics. The characteristic time in non-steady flow, such as tidal flow, and the influence of upstream supply of sediment will be discussed in subsequent subsections.

Understanding of the physical condition and mathematical modelling of water and sediment movement in rivers, estuaries and coastal waters has made much progress in recent years, but this progress has also raised many new research questions. Operational dynamic models are available for one-dimensional and two-dimensional (depth-averaged or width-averaged) simulations; see Section 3.2.

3.4.4 Conditions of transport

The dominant type of scour process depends on whether or not sediment is supplied to a scour hole from upstream. Clear-water scour occurs when bed material upstream of a potential scour hole is not in motion, generally because the flow shear stress upstream of the location is smaller than the critical shear stress of the bed material. Normally this can only hold in flume conditions and not in alluvial rivers and streams.

However, clear-water scour may also apply at scour holes where the upstream flow-induced shear stresses are larger than the critical shear stresses but not large enough to create moving bed conditions. This situation occurs, for example, in tidal rivers with clay beds. Obviously, no bed material will move if the bed is fixed or cannot erode due to a bed protection. Hence, these situations do describe clear-water scour conditions since the sediment transported towards the scour hole is essentially zero. An example occurs in tidal river branches of the Rhine-Meuse Estuary; see Figure 3.16 (Huismans et al., 2016).

Live-bed scour, in contrast, occurs where the upstream shear stress exceeds the threshold value and the bed material upstream of the scour hole moves. This means that the approach flow continuously transports sediments into a scour hole under the condition of live-bed scour. By itself, a live bed in a uniform channel will not result in a scour hole. A scour hole is only generated when the shear stress is additionally increased, which occurs, for example, at a constriction (natural or artificial, such as a groyne) or at a local obstruction (e.g. a bridge pier). In alluvial rivers and streams, the condition of live-bed scour is most common.

The generation of a scour hole differs for both conditions. The depth of a scour hole for clear-water scour approaches a limit asymptotically (Figure 3.11). When the

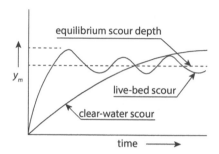

Figure 3.11 **Scour depth as a function of time.**

approach velocity is greater than the critical mean flow velocity, the upstream bed is usually covered to prevent the approaching flow from moving the bed particles. Generally, no refilling of the scour hole occurs when flow velocities have reduced after a peak for clear-water scour, due to the lack of sediment supply.

For live-bed scour, the equilibrium scour depth is smaller than that in clear-water scour conditions. In general, for the live-bed condition, the scour increases rapidly with time (Figure 3.11) and then fluctuates about a mean value in response to the bed features that pass by. The equilibrium scour depth is achieved when material is transported into the scour hole at the same rate as at which it is transported out. This equilibrium scour depth depends on the variations in the flow with depth and is approximately $y_{m,e}$ plus half the height of bed features. Equations for the equilibrium depth of a scour hole for both clear-water scour and live-bed scour are given in the next chapters.

3.5 Geotechnical aspects

3.5.1 Introduction

In or around scour holes, geotechnical processes can endanger the stability of structures. Although this book is devoted to scouring, the geotechnical aspects are connected to scour merit attention. In particular, shear failures and flow slides are important for critical slope angles. Non-homogeneous subsoils, the porosity of sand and groundwater flow play a role in the stability of slopes.

Shear failures and flow slides have occurred along banks in estuaries in the south-western part of the Netherlands. In a number of such cases, more than a million cubic metres of sand slid into the tidal channels. As a side effect, the sea dike in the vicinity was sometimes damaged or completely disappeared in the area affected. Such slides still occur from time to time, but now less failures affect the sea dikes thanks to protection works. When planning hydraulic structures in susceptible sites, the stability of these structures must be duly safeguarded.

Two major factors are generally assumed to be involved in causing shear failures and flow slides (Lindenberg & Koning, 1981). First, there is scour due to high flow velocities and turbulence along the slopes of coastal gullies in the delta region. Second, there is the porosity of the sand. It is assumed that small disturbances can initiate extensive flow slides only in loosely packed sand. With the exception of the steeper parts, the slopes that establish themselves after a flow slide are nearly always milder than 1V:10H and in some cases only as mild as 1V:20H to 1V:25H. However, the effects of ground-water flow and the non-homogeneity of the soil play a role too.

Flow slides can occur in two ways: (1) liquefaction flow slide or static soil liquefaction and (2) breach flow slide or breaching. Here, mainly liquefaction flow slide will be discussed.

Both types of flow slides involve soil mechanical and hydraulic features. They result in a flowing sand–water mixture, exhibiting the same post-event failure profile morphology and characterised by a gentle slope. Therefore, it is often not clear in the analysis of historical flow slides to what extent each of the two phenomena was responsible.

Liquefaction flow slide entails the sudden loss of strength of loosely packed saturated sand or silt, resulting in a sudden collapse of the sand body. Contrary to "ordinary" slope failure, in which the instable soil mass slides along a clear rupture surface while staying more or less intact, the unstable mass of sand (or silt) of a liquefaction flow slide moves as a laminar flow like a viscous fluid.

Generally, the following conditions are required for static liquefaction in an underwater slope: (1) a sufficiently large zone of loosely packed, non-lithified and water-saturated sand or silt should be present; (2) the stress state of the loosely packed sand elements should be close to the metastability point (i.e. a stable state in which both mean stress and shear stress should be sufficiently large, which is only the case in a sufficiently high and steep slope); and (3) the liquefaction should be triggered, for example, by a (small) load change.

Unlike liquefaction and "ordinary" shear slope failure, breaching only takes place at the sand surface. A local steep part of the slope, called "breach", retrogresses upslope and generates a turbulent sand-water mixture flow downslope over the sand surface (Figure 3.12). A local steep part of the slope, called "breach", retrogresses upslope and generates a turbulent sand–water mixture flow downslope over the sand surface. The velocity and discharge of this mixture will grow by erosion of the sand surface and entrainment of ambient water if (a) the initial perturbation generates a sufficiently high flow velocity carrying enough sand, and (b) the local slope is steep enough. Since the retrogression velocity of the initial breach is relatively small, a breaching flow slide generally takes much more time (several hours) than a liquefaction flow slide (several minutes).

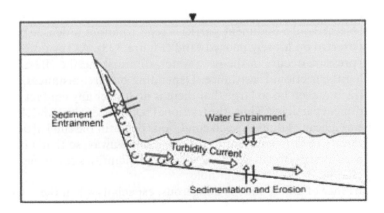

Figure 3.12 Conceptual sketch of the governing physics of a breaching process

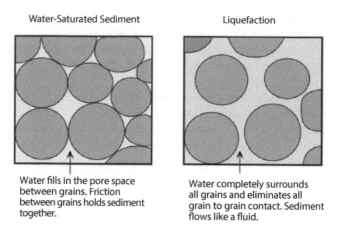

Water-Saturated Sediment

Liquefaction

Water fills in the pore space between grains. Friction between grains holds sediment together.

Water completely surrounds all grains and eliminates all grain to grain contact. Sediment flows like a fluid.

Figure 3.13 **Sediment with and without frictional resistance.**

Generally, the following conditions are required for liquefaction flow slide in an underwater slope: (1) the presence of a sufficiently large zone of fine sand or silt; (2) a sufficiently high and steep slope; and (3) the presence of a trigger, for example, scour or a local slope instability yielding a small but steep slope section (breach). More information is given by de Groot et al. (2009), Mastbergen and van den Berg (2003), and van den Ham et al. (2014) (Figure 3.12).

We first discuss the influence of the porosity of sand on scouring in Section 3.5.2. Then, the effects of groundwater flow is treated in Section 3.5.3. Section 3.5.4 pays attention to non-homogeneous soils. Section 3.5.5 presents empirical criteria for slopes, whereas Section 3.5.6 discusses the failure length or the critical length of the bed protection.

3.5.2 Liquefaction

Sand is considered to be loosely packed if the porosity (percentage of voids) of the sand is higher than a critical value, which is approximately 40% in the south-western part of the Netherlands. Sediment particles tend to adopt a denser packing if shear stresses are exerted on loosely packed sand (Figure 3.13). As the pores are filled with water, over-pressure occurs in the pore water, diminishing the effective stresses and thus reducing the frictional resistance. Depending on circumstances, the increase in over-pressure may even be so large that there is no longer any contact between the individual sediment particles. Thus, the frictional resistance is lost and the sand behaves more or less as a heavy liquid. If such conditions occur in the subsoil adjacent to steep slopes, a quantity of sand adjacent to the slope slides away, so that a part of the slope higher up loses its support. In this way, very large quantities can liquefy and flow into the deeper channel or into the scour hole.

The occurrence of a flow slide is dangerous, especially when the affected soil mass extends up to the bed protection of a hydraulic structure. In such cases, the flow slide initiated by the upstream scour slope can undermine the structure by progressive failure under the bed protection, leading to a major structural failure. Therefore, it is essential that the porosity or the critical density in situ is known.

3.5.3 Effects of groundwater flow

In the subsiding delta of the Netherlands, high-discharge water levels in embanked rivers are higher than water levels in the surrounding polders. Water then flows to the polders through permeable layers under the dikes. This creates an upward ground-water flow in the polder. Similar effects may occur, for example, at canal dikes or secondary polder dikes. Rainfall may saturate such dikes and cause water to flow out to the canal bottom or the polder water, transporting subsoil material. When this oc-curs for an elongated period of time, this may endanger the embankment, or the bank protection may slide. A flow of water to the canal may result in bursting of the canal bottom if this bottom consists of a clay lining. This phenomenon can also occur at locks. Therefore, the concrete floor or impermeable bed protection of a lock has to be designed against floating due to upward groundwater flows.

Erosion of soil material below hydraulic structures is called internal erosion. The process of internal erosion decreases soil stability and eventually endangers the sta-bility of the hydraulic structure by triggering sliding or scouring. It is important to be aware of this type of erosion. Various types can be distinguished:

1. piping (or backward erosion)
2. contact erosion
3. suffusion.

The term piping is internationally used for all types of internal erosion mentioned above, but in the Netherlands piping is nearly always related to backward erosion; see Figure 3.14. During piping, concentrated seepage flow transports the soil material in a sand layer below a hydraulic structure. The flow creates an open pipe only a few sand grains wide. The failure process consists of a number of partial mechanisms: formation of a well, bursting (possibly), pipe-shaping, enlargement of the pipe and underflow of the structure.

Contact-erosion may occur when coarse materials such as gravel overlays and fine materials such as sand are in a granular filter structure. The strong flow in the layer with the large material may erode the fine material. For this type of erosion in bed protections, downstream hydraulic structures should be prevented. Design rules can be found in Hoffmans (2012) and CUR manual 233 (CUR 233, 2010). Figure 3.15 shows that sand erodes because the critical flow velocity in the filter layer is exceeded.

Suffusion is that the fine particles in a layer with the coarse material are washed away while the larger grains keep the grain skeleton intact; see Figure 3.16. Suffusion is only possible in mixtures with a broad non-uniform grain size distribution that allows small grains to move through the openings in the skeleton of the larger grains. Dutch sand is relatively small and uniform, and subsequently, there is hardly any risk of suffusion. Gravel and larger rock mixtures with a strongly non-uniform distribution should be evaluated on the suffusion criterion; see ICOLD-bulletin (2015) or the CUR 233 manual on granular filters.

For detailed information, the reader is referred to the CUR publications CUR 233 (2010), CUR 161 (1993) and SBRCURnet Kennispaper (CUR, 2015).

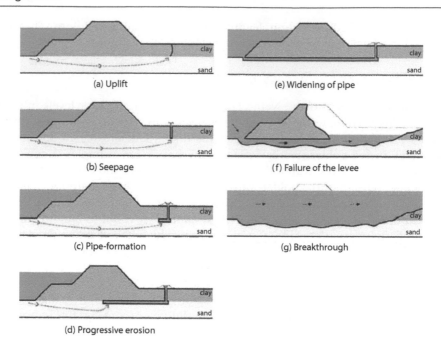

(a) Uplift

(e) Widening of pipe

(b) Seepage

(f) Failure of the levee

(c) Pipe-formation

(g) Breakthrough

(d) Progressive erosion

Figure 3.14 **Progressive erosion process.**

Figure 3.15 **Steady flow parallel to the interface (horizontal filter; flow direction from left to right).**

3.5.4 Non-homogeneous subsoils

Scour development in non-homogeneous sub-soils (or non-uniform soils or heterogenous soils) is still an area of great uncertainty and remains a challenge for designing structurally efficient and effective foundations in (tidal) river environments. Scour risk

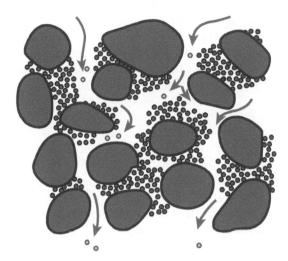

Figure 3.16 Detachment, transport and filtration of the fine solid fraction of a soil under the action of water seepage during internal erosion by suffusion.

Figure 3.17 Bird eye view of a scour hole in the Spui near Nieuw-Beijerland. (Courtesy: Kees Sloff.)

in cohesive soils is made more uncertain by effects such as weathering and time scale to scour. The holes may form a potential risk for the stability of dikes, bridges, tunnels and buildings (Figure 3.17). Hence, the geology and geomorphology of risk sites need to be studied to determine the potential for long-term bed elevation changes at these locations.

For large hydraulic structures and surrounding riverbanks, there is a limit to the amount of detailed geotechnical information that can be collected as part of the project. Therefore, reliance on data such as undrained shear strength, derived from cone penetration tests, supplemented with borehole data, collected at a limited number of sites across the structures and dikes, and laboratory analysis of soil samples become the principal source of geotechnical information.

Heterogeneity of the subsoil in the Rhine-Meuse Delta in the Netherlands causes differences in erosive capacity, which results in the formation of deep scour holes (Sloff et al., 2013, see also Figure 3.17). A scour hole can mostly be the result of two conditions: local changes in hydrodynamic conditions or local changes in erosion resistance of the bed. A distinct method to predict the development of scour holes in non-homogenous sub-soil is not found yet; more information on this subject can be found in Bom (2017).

3.5.5 Upstream and side slopes

Flow slides usually have more serious consequences for hydraulic structures than shear failures. The length of the expected damage of the bed protection caused by progressive shear failure can be determined by using standard methods, as developed, for example, by Bishop (1954) and Fellenius (1947). However, the mechanism of the two types of flow slides is still not fully understood, so it is difficult to predict the damage and rate of the backward erosion accurately.

Although little information is available concerning instabilities, some rough design criteria based on a two-dimensional approach have been established. Figure 3.18 shows

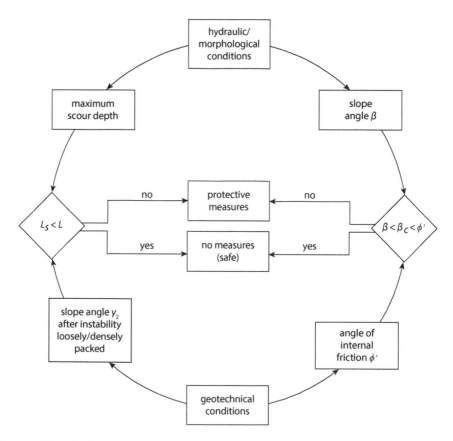

Figure 3.18 Relation between loading and strength parameters.

Table 3.2 Upper values of upstream scour slopes for soil mechanical instabilities (Silvis, 1988)

Bulk density of fine sand	Angle of internal friction ϕ'	Critical slope angle β_c	Phenomenon
Very loose	30° (IV:1.7H)	24° (IV:2.25H)	Flow slide
Loose	33° (IV:1.5H)	27° (IV:2.00H)	Flow slide
Dense	37° (IV:1.3H)	30° (IV:1.75H)	Shear failure
Very dense	40° (IV:1.2H)	34° (IV:1.50H)	Shear failure

Notes
1 The critical values are given for which the probability of geotechnical instabilities approaches 1.
2 The critical slopes for a flow slide are too steep; a slide may occur at smaller slopes, depending on the slope height and the density (with lower density and higher slope, the critical slope will be decreased).

the principle of design with respect to the minimum length of the bed protection and the upstream scour slope angle (β). When the length of the bed protection is too short ($L_s \geq L$), or when the slope angle β exceeds the critical upstream scour slope, additional measures have to be carried out to ensure the safety of the hydraulic structure. The right-hand side of Figure 3.18 implicitly shows that if a slope becomes too steep local instabilities will occur and countermeasures are required (e.g. rock dumping). Note that Figure 3.18 incorporates both geotechnical and morphological aspects.

Based on field measurements (Wilderom, 1979) regarding dike failures in the south-western part of the Netherlands, Silvis (1988) determined some empirical criteria for critical slopes (Table 3.2). These criteria regard critical slopes or inclinations for which the probability of geotechnical instabilities approaches 1. However, these criteria do not include the retarding effect of the bed protection, so that the upstream scour slope might be somewhat steeper if the bed upstream of the scour hole is protected. As the lower value (conservative) for loose to very loose sand, a critical slope gentler than 1V:4H should be introduced.

In addition to slope criteria, also criteria related to the relative density D_r which defines the state of compactness are important:

- $D_r < 0.33$: very sensitive for a sliding
- $0.33 < D_r < 0.67$: sensitive
- $D_r > 0.67$: not sensitive

$$D_r = \frac{e_{max} - e}{e_{max} - e_{min}} \tag{3.17}$$

in which
  D_r = relative soil density (-),
  e = actual void ratio (-), $e = \dfrac{V_v}{V_s}$
  e_{max} = maximum void ratio (-)
  e_{min} = minimum void ratio (-)

V_v = volume of voids
V_s = volume of solids

The geotechnical relative density D_r can be determined via the CPT (Cone Penetration Test) as defined by Baldi et al. (1982). According to Sowers (see HEC-18), the classification of D_r reads as follows:

Table 3.3 Relative density as function of soil classification

Soil classification	D_r
Loose	0–0.5
Firm	0.5–0.7
Dense	0.7–0.9
Very dense	0.9–1.0

The side slope can be expressed by the angle of repose because the angle of repose is an upper limit for the side slope in a two-dimensional geometry. In a local scour hole, the lateral support at a concave three-dimensional surface may allow a stable slope steeper than the angle of repose. However, this effect has been observed only in sand with some cohesion caused by clay particles, for example, in the prototype tests in the Brouwers Sluice but not in pure non-cohesive sand used in laboratory tests. The angle of repose depends on the type of soil and also on the compaction of the soil. For some characteristic values, see Table 3.4.

Analysis of geotechnical scour events in 2012 at the Eastern Scheldt storm surge barrier (see also Section 6.6.4) made clear that the criteria of Silvis are also applicable for side slopes and not for upstream slopes only. On side slopes also flow slides may occur at slopes less steep than mentioned in Table 3.4 (and again in combination with height and density). Side slope failures can also be triggered by slope failure of the upstream slope and vice versa.

Table 3.4 The angle of repose for different soils

	Soil type	Angle of repose ϕ
Coarse sand	Compact	45°
Sand and gravel	Firm	38°
	Loose	32°
Medium sand	Compact	40°
	Firm	34°
	Loose	30°
Fine sand, silty sand, sandy silt	Compact	30–34°
	Firm	28–30°
	Loose	26–28°
Clay (saturated)	Medium	10–20°
	Soft	0–20°

3.5.6 Failure length

Wilderom (1979) reported that the failure length due to instabilities of the subsoil depends strongly on the storage capacity of the channel in front of the foreland. When the bed of the scour hole downstream of the point of reattachment is assumed to be horizontal, the failure length can be approximated with the following geometrical equation presented by Silvis (1988):

$$L_s = y_d \left(1/2 \frac{y_d}{y_m} - 1 \right) (\cot \gamma_2 - \cot \gamma_1) + 1/2 \, y_m (\cot \gamma_2 - \cot \beta_a) \qquad (3.18)$$

in which
 L_s = failure length (m)
 y_d = depth after a sliding (m)
 y_m = maximum scour depth (m)
 β_a = average slope angle before instability (°)
 γ_1 = sliding erosion slope angle after instability (°)
 γ_2 = sliding deposit slope angle after instability (°)

Equation (3.18) has been based on a two-dimensional storage model in which the deepening (volume of liquefied sand) equals the deposit $O_1 = O_2$ (Figure 3.19).

Table 3.5 gives an overview of mean and extreme values of the parameters in the aforementioned storage model. These results are based on approximately 200 instabilities caused by shear failures or flow slides (Wilderom, 1979). Applying default values, the computed failure length must be considered as a first estimate ($L_s \approx 2y_m$) since specific information regarding the subsoil (e.g. porosity of sand, clay layers) has not been taken into account.

A rough but conservative failure length is obtained if $y_d = 0$. With this assumption, Equation (3.18) gives an upper limit and simplifies into:

$$L_s = 1/2 \, y_m (\cot \gamma_2 - \cot \beta_a) \qquad (3.19)$$

The failure length is reduced when the volume of the scour hole limits the storage capacity of the flow slide. For a triangular scour hole, the failure length can be given by the generally applicable geometrical formula (Silvis, 1988):

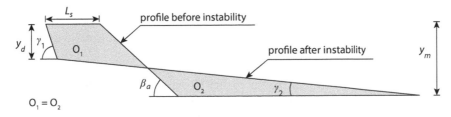

Figure 3.19 Two-dimensional schematisation of a flow slide.

Table 3.5 Statistical parameters of shear failures and flow slides

Storage parameters	Flow slides and shear failures (Silvis, 1988) – mean values	Shear failures (Konter et al., 1992) – extreme values		Flow slides (Konter et al., 1992) – extreme values	
$\cot \beta_a$	3.7	4^a	2^b	4^a	2^b
$\cot \gamma_1$	2.7	2	2	2	2
$\cot \gamma_2$	16.3	8	8	20	20
$\cot \delta$	40				
y_d/y_m	0.43	0	0	0	0
L_s/y_m (no storage)	1.7	2.0	3.0	8.0	9.0
L_s/y_m (storage)	1.5	2.0	2.9	7.5	8.3

[a]   Protected slope.
[b]   Unprotected slope.

$$L_s = \sqrt{(\cot \gamma_2 - \cot \delta)\left(a y_m^2 + b y_d^2\right)} - a y_m - b y_d \qquad (3.20)$$

in which
$a = \cot \beta_a + \cot \delta$ (–)
$b = \cot \gamma_2 - \cot \gamma_1$ (–)
δ = slope angle downstream of the deepest point of the scour hole (°)

Note that both equations are conservative with respect to the stability of a protection if only one flow slide occurs. However, after a sliding, a new scour hole will develop and a second or third sliding cannot be excluded. Hence, it is important to monitor the situation after a sliding to prevent a second sliding.

De Graauw (1981) derived a similar equation from a less accurate schematisation. The execution of a thorough in-situ soil investigation may lead to the conclusion that the subsoil consists of dense sand. In such cases, the value of γ_2 can be decreased significantly, leading to a shorter bed protection length. It should be noted that the critical failure length depends strongly on γ_2 (slope angle after instability). At present, no relations are available that relate γ_2 accurately to soil parameters. The following rough values are recommended: flow slides $15 < \cot \gamma_2 < 20$ and shear failures $6 < \cot \gamma_2 < 8$.

The above findings suggest that in the case of dense sand, the value of γ_2 can be chosen smaller resulting in a "shorter bed protection length". However, it has never been proven that for dense sand, the slope becomes steeper. Clearly, there is a relation between the grain size and γ_2: fine sand settles on a less steep slope than coarse sand, but an equation has never been derived (although computations with the Deltares model HMBreach support this).

Finally, CUR 113 (2008) provides simple occurrence criteria for both flow slides and breach slides (Figure 3.18). Also, Voorschrift Toetsen op Veiligheid (VTV, 2016) provides criteria (per km per year) that are comparable with the ones in CUR113 and that have been based on flow slides in the Dutch province of Zeeland. Even the probability distribution of failure length L_s can be computed by using a slightly modified Silvis equation. In addition, VTV (2016) provides criteria on the rate of risk, see both manuals for more information.

3.6 Examples

3.6.1 Introduction

Section 3.6.2 shows the computation of general scour at a constriction in a river due to a bridge crossing. The example only shows the scour as a result of the narrower channel and does not include local scour at the bridge piers and the approach abutments. In Section 3.6.3, the critical slope angles and failure lengths downstream of a hydraulic structure are computed to show the relevance of taking into account sufficient length of the bed protection and values figures for the internal angle of friction.

3.6.2 Constriction scour

Constriction scour can be determined with Equation (3.1):

$$\frac{y_{m,e} + h_0}{h_0} = \frac{1}{(1-m)^{\beta}} \tag{3.1}$$

Assume the following input (see Figure 3.4): upstream river width $B_1 = 98.2\,\text{m}$ and upstream flow depth $h_0 = 2.62\,\text{m}$. The bridge has three piers with $b =$ pier diameter $= 0.38\,\text{m}$. The distance between the bridge abutments is $37.2\,\text{m}$ which results in an effective width of the river at the constriction $B_2 = 37.2 - 3 \times 0.38 = 36.06\,\text{m}$.

The constriction ratio now follows with $m = 1 - B_2/B_1 = 1 - 36.06/98.2 = 0.63$.

Substituting these values in Equation (3.1) gives the result for the equilibrium scour depth: $y_{m,e} = 2.5 - 3.2\,\text{m}$ (for $\beta = 0.67 - 0.8$).

3.6.3 Critical slope angles and failure lengths

A hydraulic structure is built on loosely packed sand. The bed downstream of the hydraulic structure is protected against current and eddies. The upstream scour slope is about $\cot \beta_a = 4.0$ and protected with rock. In addition, to ensure the safety of the structure, it is protected against the occurrence of flow slides and of shear failures with an extreme probability. The angle of internal friction amounts to $\phi' = 33°$.

a. What is the critical slope angle according to Silvis (1988)?
b. What is the minimum length of the bed protection?
c. What would be the failure length if the subsoils consisted of both loosely and very densely packed sand? Up to $5\,\text{m}$ below the bed protection, the soil is loosely packed.
d. What would be the failure length if the scour slope downstream of the deepest point of the scour hole was about $\cot \delta = 25$? Other data are $\cot \gamma_1 = 2.0$, $\cot \gamma_2 = 25$, $y_d = 0.2 y_m$.

e. What would be the critical slope angle if the subsoil consists of densely packed sand and what would be the corresponding failure length following Silvis (1988)? ($\phi' = 38°$).

Solution:

a. $\beta_c = 27°$

b. $L_s = 8.0 y_m$ (no storage capacity) or $L_s = 7.5 y_m$ (storage capacity)

c. $L_s = 8 y_m$

d. $L_s = \sqrt{\left[(\cot \gamma_2 + \cot \delta) \times (a \times y_m^2 + b \times y_d^2) \right]} - a \times y_m - b \times y_d$

   $a = \cot \beta_a + \cot \delta = 4 + 25 = 29$

   $b = \cot \gamma_2 - \cot \gamma_1 = 25 - 2 = 23$

   $L_s = \sqrt{\left[(25 + 25) \times (29 \times y_m^2 + 23 \times (0.2 \times y_m)^2) \right]} - 29 \times y_m - 23 \times 0.2 \times y_m$

   $= 38.7 \times y_m - 29 \times y_m - 4.6 \times y_m = 5.1 y_m$

e. $\beta_c = 31°$ and $L_s \approx 2.0 y_m$

Chapter 4

Initiation of motion

4.1 Introduction

Shields (1936) presented the first treatise on initial bed grain instability, referring to Prandtl's and von Kármán's concepts of boundary-layer flow that are mentioned in the bibliography of this manual. He described the problem using the following parameters: the fluid density, the sediment density, the kinematic viscosity, the grain size and the bed shear stress. When the flow velocity over a bed of non-cohesive materials has increased sufficiently, individual grains begin to move in an intermittent and random fashion. Bed instability results from the interaction between two stochastic variables. The first stochastic variable is the instantaneous bed shear stress exerted by the flow. The second one is the instantaneous critical shear stress for bringing a grain on the bed surface into motion. Due to the random shape, weight and placement of the individual grains, these critical shear stresses will have a probability distribution. The grain becomes unstable if the instantaneous bed shear stress exceeds the critical shear stress. The probability that the instantaneous bed shear stress is greater than a characteristic critical shear stress is a measure of the transport of sediments. The mechanics of interactions between fluid flow and sediment is the subject of numerous papers and a number of textbooks, for example Bogardi (1974), Graf (1971), Raudkivi (1993), van Rijn (1993), Yalin (1972, 1992) and Schiereck and Verhagen (2012).

The physio-chemical properties of cohesive sediments play a significant role in the resistance of cohesive sediments to currents. These properties depend strongly on the granulometric, mineralogical and chemical characteristics of the sediment involved. Until now, direct quantitative relations between the physio-chemical properties and the erosion rate have not been established. Nevertheless, design engineers require information to predict scour in cohesive sediments because these soils are widespread natural sedimentary deposits. In Section 4.2, flow and turbulence are treated in an introductory way as well as other flow characteristics. The governing parameters for the erosion of non-cohesive and cohesive sediments are discussed in Sections 4.3 and 4.4, respectively. Finally, some examples are given in Section 4.5.

4.2 Flow and turbulence characteristics

4.2.1 Introduction

The bed turbulence (defined as the standard deviation of the instantaneous bed shear stress) along with the mean bed shear stress determine the bed load transport of a given bed grain size fraction. For uniform flow, the ratio between the bed turbulence and the bed shear stress is approximately constant (Compte-Bellot, 1963), whereas for non-uniform flow, the bed turbulence and the bed shear stress are strongly influenced by turbulent energy generated in the mixing layer and a combination of vortices with both horizontal and vertical axis. Hoffmans (1992) and Hoffmans and Booij (1993a, b) have shown that in a two-dimensional scour hole, the bed turbulence can be represented by a combination of the turbulent energy generated at the bed and the turbulent energy from the mixing layer.

Several researchers have investigated the influence of turbulence on the bed load. As given by Kalinske (1947) and Einstein (1950), the instantaneous velocity varies according to a Gaussian distribution. Kalinske's idea was extended by van Rijn (1985), who postulated an instantaneous transport parameter in which the instantaneous bed shear stress is normally distributed. The weakness of these classical stochastic models is that they do not incorporate more modern understanding of turbulence. As a result, they are not based on the mechanics of turbulence but use only the continuity equation for sediment transport and the parametric probability density functions. For example, the measurements of Lu and Willmarth (1973) show that for uniform flow, the influence of sweeps and ejections is not included in the Gaussian distribution. Sweeps, which are directed towards the bed, and ejections, which are moving away from the bed, contribute most to the turbulent shear stresses (Figure 4.1). Under non-uniform flow conditions, there is no clear relation between the instantaneous sediment transport and the instantaneous bed shear stress. Near-bed measurements of turbulent correlations (Reynolds stresses) are estimates of momentum flux but are only related to the force acting on the bed when it is averaged over a long period of time.

The relation between instantaneous products of velocity components and the instantaneous force on a sediment particle is not fully understood. Nevertheless, some general premises have been used to model the two important design parameters of the scour process: maximum scour depth and upstream scour slope. According to Breusers (1966, 1967), the sediment transport in a scour hole is related to the difference between a maximum and a critical flow velocity raised to a power.

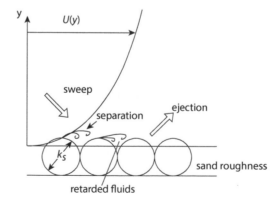

Figure 4.1 Schematic descriptions of turbulent flow over smooth and rough beds (Nezu & Nakagawa, 1993).

The maximum flow velocity is a function of the mean velocity and the turbulent kinetic energy defined as:

$$U_{max} = U\left(1+3r_0\right) = U\left(1+3\frac{\sqrt{k}}{U}\right) = U + 3\sqrt{k} \tag{4.1}$$

with

$$k = \frac{1}{2}\left(\sigma_u^2 + \sigma_v^2 + \sigma_w^2\right) = \left(r_0 U\right)^2 \approx \sigma_u^2 \tag{4.2}$$

in which
k = turbulent kinetic energy (m^2/s^2)
r_0 = relative turbulence intensity (–)
U = time-averaged velocity (m/s)
U_{max} = maximum velocity (m/s)
σ_u = standard deviation of instantaneous velocity in the streamwise direction (m/s)
σ_v = standard deviation of instantaneous velocity in the transverse direction (m/s)
σ_w = standard deviation of instantaneous velocity in the vertical direction (m/s)

Usually Equations (4.1) and (4.2) describe local values. However, for predicting sediment transport, and scouring they refer to depth-averaged parameters. For uniform flow conditions, r_0 varies from 0.05 (hydraulically smooth) to 0.15 (hydraulically rough). In the deepest part of the scour hole, the relative turbulence intensity $r_{0,m}$ is:

$$r_{0,m} = \frac{\sqrt{k_m}}{U_m} \tag{4.3}$$

in which
k_m = mean turbulent kinetic energy where scour depth is at maximum (m^2/s^2)
U_m = depth-averaged flow velocity where scour depth is at maximum (m/s)

For non-uniform flow, the relative turbulence intensity $r_{0,m}$ lies in the range of 0.15–0.60. Directly downstream of a hydraulic jump or a recirculation zone near the bed, the representative turbulent kinetic energy in the mixing layer and the depth-averaged value can be written as (Van Mierlo & De Ruiter, 1988):

$$k_{max} = C_k U_0^2 \tag{4.4}$$

$$k_m = \frac{1}{2}k_{max} \tag{4.5}$$

in which
C_k = constant that ranges from 0.030 (for scour slopes less steep than 1V:3H) to 0.045 (for backward facing step) (Figure 4.2)
k_{max} = maximum turbulent kinetic energy in the mixing layer (m^2/s^2)
U_0 = depth-averaged flow velocity upstream of the scour hole (m/s)

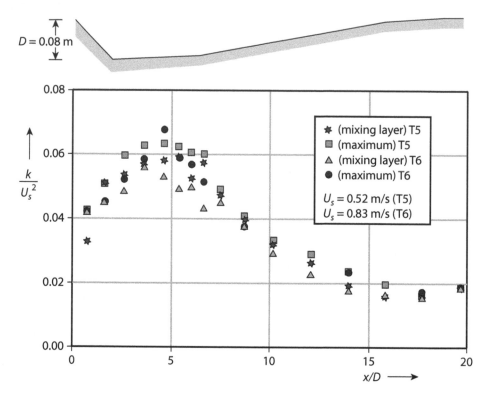

Figure 4.2 k_{max} as a function of x/D with D the step height. k_{max} increases about linearly in the mixing layer and reaches its maximum value where the flow depth is near its maximum value. Then, k_{max} decreases to an equilibrium value. (Measurements obtained from Van Mierlo & De Ruiter, 1988.)

4.2.2 Sills

Downstream of a sill, the relative turbulence intensity depends strongly on the height of the sill (relative to the flow depth), the distance from the sill and the roughness of the bed. A mixing layer and a recirculation zone develop. Downstream of the point of reattachment, a new boundary layer starts developing (see Figure 4.3). Above the boundary layer, i.e. in the relaxation area, the turbulence decreases gradually to an equilibrium value. This value is reached when the thickness of the mixing layer equals the water depth. With the increasing length of the bed protection, the dimensions of the scour hole decrease because the load decreases. However, there is a minimum value determined by the bed roughness and the transition at the sill. The decrease of the turbulence is described generally with the turbulent kinetic energy but can also be modelled with less known parameters, such as turbulent viscosity, dissipation or shear stresses (Launder & Spalding, 1972). At a distance of 20–50 times the flow depth downstream of the reattachment point, r_0 is almost equal to the uniform turbulence level.

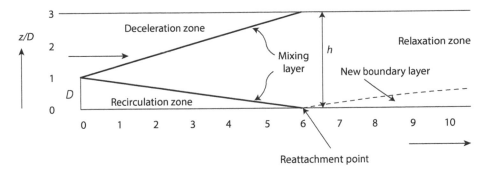

Figure 4.3 Schematisation of hydrodynamic zones downstream of a sill.

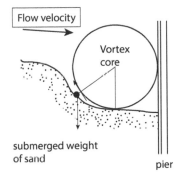

Figure 4.4 Eddy with horizontal axis (Tsujimoto, 1988).

4.2.3 Bridge piers and abutments

It is possible to estimate $r_{0,m}$ not only downstream of sills but also at piers and abutments where scouring is expected. Figure 4.4 shows a vortex upstream of a pier with a constant rotational direction. Above the recirculation zone, a mixing layer occurs, yielding higher bed turbulence intensities due to horse shoe vortices; see also Chapters 7 and 8. Hawkswood and King (2016) presented local turbulence intensities (not depth-averaged intensities); see Table 4.1.

In the case of abutments and groynes, the flow velocity increases at the constriction resulting in a gradual scouring. Downstream of the groynes, the flow velocity decreases. In this deceleration area, a mixing layer develops that generates turbulent kinetic energy. The mixing layer is characteristic for the equilibrium scour depth at abutments; see Figure 4.5.

For sills, the damping of turbulence can be estimated with simple formulas based on geometrical dimensions such as sill height and length of bed protection (see Equation 6.11). Such simple formulas for the damping of turbulence are not available for bridge piers and abutments. In the Netherlands, bridge piers and the heads of abutments or groynes are protected, which in fact means that the scour process moves

Table 4.1 Local turbulence intensities at bridges (Hawkswood and King, 2016)

Turbulence intensity r	Turbulence conditions at bridges
0.12	Normal river beds
0.20	Upstream edges to beds
0.25	Low bridge constriction (variable bottom and slope configurations)
0.35	Typical bridge constriction
0.50	High bridge constriction (wide piers)

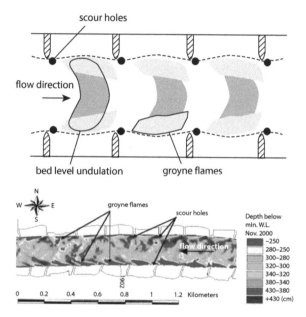

Figure 4.5 Groynes-induced morphology: definition sketch (upper panel) and example from River Waal near Druten, the Netherlands (lower panel).

away from the structure in such a way that the geotechnical stability will never be endangered. Therefore, it is recommended to develop such simple formulas also for bridge piers and groynes.

In case of bridge piers, abutments and groynes, the angle of flow attack plays a role. The question arises how to incorporate the angle of flow attack in the design of the bed protection. This will be shown in a case in Chapter 9.

4.2.4 Indicative values of flow velocity and turbulence

Nezu and Nakagawa (1993) have surveyed a broad spectrum of technical and scientific literature on the subject of turbulence in open-channel flow. The text of their state-of-the-art monograph is up-to-date and forms an invaluable source of information. The authors propose an interaction mechanism between the coherent structure of turbulence and the initiation of the development of bed forms such as ripples, dunes

Table 4.2 Indicative values of depth-averaged flow velocity

Type of flow	U_0 (m/s)
River/outer bend river	2–3
Ship lock/sluice	1–2
Tidal river (downstream of storm surge barrier)	2–3
Culvert (outflow opening) (subcritical flow)	1–3
Culvert (outflow opening) (supercritical flow)	3–8
Return current of ship	1
Propeller jet of ship (manoeuvring ship)	2–5

Table 4.3 Indicative values of r_0 (Hoffmans, 2012)

r_0	Type of Flow	Turbulence intensity	Hydraulic roughness	Remarks
0	Laminar flow	No turbulence		
>0	Turbulent flow			
<0.08	Uniform flow	Minor turbulence	Smooth	
0.08–0.15	Uniform flow	Normal turbulence	Rough	Channel/river flow
0.15–0.20	Uniform flow	High turbulence	↓	Channel/river flow
0.20–0.30	Uniform flow	Very high turbulence		Steep channel with limited flow depth (0.01–0.1 m)
0.30–0.60	Uniform flow	Extreme high turbulence		Very steep channel with limited flow depth
0.15–0.20	Non-uniform flow	High turbulence		(sills, bridge piers, abutments)
0.20–0.60	Non-uniform flow	Very high turbulence		Below hydraulic jumps, sharp outer bends, mixing layers (propellers)

and anti-dunes. Further development of physics-based models and formulae incorporating these coherent structures are necessary to further progress in this field.

Hoffmans (2012) makes a distinction between average hydraulic load (average flow velocity) and the range, i.e. the fluctuating flow velocity or turbulence. Tables 4.2 and 4.3 provide information about upper limits of flow velocities and relative turbulence for different conditions on the basis of expert judgement. The turbulence coefficient strongly affects the rate of scouring and is a function of the depth-averaged relative turbulence intensity.

4.3 Non-cohesive sediments

4.3.1 Introduction

In the literature, numerous relations can be found for the critical flow velocity for particle movement (e.g. Brahms, 1767; Izbash, 1932; Izbash & Khaldre, 1970). One of the oldest curves to predict the behaviour of bed materials is the Hjulström (1935) curve presented in Figure 4.6. Actually, it is a wrong concept for the determination of

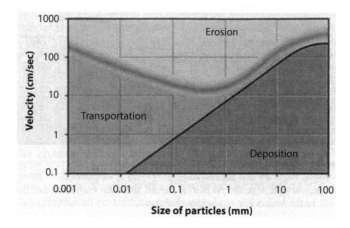

Figure 4.6 Hjulström diagram.
Note: Chart is based on limited number of data. Use with caution.

scour or sedimentation in a flow because in reality, no critical flow velocity exists for sediment deposition. Moreover, the threshold between transportation without erosion and transportation with erosion does not depend on flow velocity but on gradients in sediment transport capacity. The Hjulström curve is of historical interest but is no longer considered to be of any use for the mechanics of sediments in flowing water.

Briaud (2019) published Figure 4.7 that gives a first idea about the sensitivity for erosion of various materials expressed as erosion rate as function of the flow velocity and the erodibility of the soil material. More information can be found in Shafii et al. (2016).

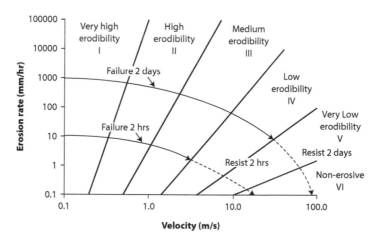

Figure 4.7 Erosion rate as a function of the flow velocity for various soil materials (Briaud, 2019).
Note: I = Very High Erodibility: fine sand, silt; II = High Erodibility: medium sand, silt; III = Medium Erodibility: coarse sand, fine gravel, clay; IV = Low Erodibility: cobbles, coarse gravel, clay; V = Very Low Erodibility: riprap; VI = Non-Erosive: rock

In the next section, approaches are presented regarding critical conditions for non-cohesive materials.

4.3.2 Shields diagram

In 1936, Shields published his criterion for the initiation of movement of uniform granular materials on a flat bed. The experimental data used by Shields were mainly obtained by extrapolating curves of sediment transport versus applied shear stress towards the zero-transport condition. Originally, the data points were plotted by Shields and the curve (averaged critical value), constituting the 'Shields diagram', as usually quoted (Neill, 1968), was drawn by Rouse (dashed line in Figure 4.8).

Actually, because of the non-uniform distribution of the mixtures and the effects of grain imbrication (i.e. the preferred orientation of natural sands and gravel particles under certain conditions of transport), Shields did not draw a single curve but a broad belt of curves.

The critical bed shear stress (or critical mobility parameter) can be obtained graphically, directly from the modified Shields diagram (Figures 4.8 and 4.9) or by using expressions that fit the Shields diagram.

Figure 4.8 essentially shows seven criteria for initiation of motion. Breusers added these criteria based on his investigations between 1960 and 1980. He reported that around 1920 or 1930, four qualitative criteria were distinguished:

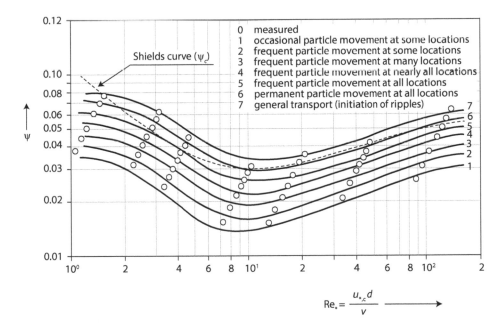

$$Re_* = \frac{u_{*,c} d}{\nu}$$

Figure 4.8 **Shields diagram; Shields parameter as function of the particle Reynolds number (critical Shields parameter (dashed line) and seven lines for different criteria for initiation of motion).**

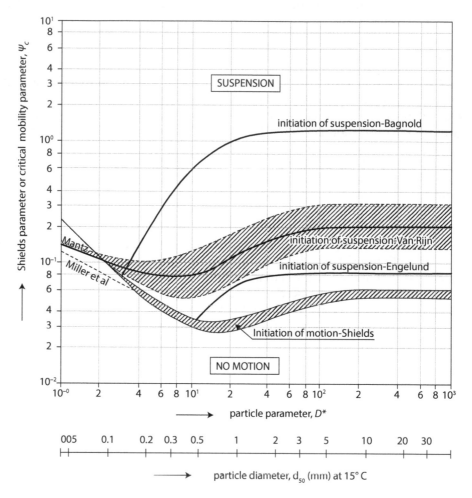

Figure 4.9 **Initiation of motion and suspension for a current over a plane bed (van Rijn, 2007).**

1. No transport of bed material – all bed material is in absolute rest
2. Limited transport of bed material – small, countable numbers of grains are moving, mostly locally
3. Average transport of bed material – number of moving grains cannot be counted anymore, however no significant transport
4. General transport of bed material – all grains are moving, transport of grains

The Shields parameter Ψ can be related to the dimensionless transport rate ϕ which is a function of the sediment transport. Paintal (1971) does this for small transports with Ψ values in the range of 0.04–0.06. Further information can be found in sediment transport manuals, for example Van Rijn (1993, 2007).

4.3.3 Design approaches

The Shields diagram has been applied numerous times for the prediction of initiation of motion. However, the critical Shields parameter itself is neither a hydraulic load nor a strength parameter. It represents the ratio between the load and strength.

The critical bed shear stress (τ_c) is defined as follows:

$$\tau_c = \rho u_{*,c}^2 \tag{4.6}$$

According to Shields (1936), the critical Shields parameter is:

$$\Psi_c = \frac{u_{*,c}^2}{\Delta g d} \tag{4.7}$$

in which
Ψ_c = critical Shields parameter (–)
d = particle diameter (m) ($d = d_{50}$ is median grain size)
g = acceleration of gravity, $g = 9.8 \, \text{m/s}^2$
$u_{*,c}$ = critical bed shear velocity (m/s)
Δ = relative density (–)
ρ = fluid density (kg/m^3)

Erosion starts if the Shields parameter exceeds a critical value, thus:

$$\Psi \geq \Psi_c$$

With an increasing value of the Shields parameter, the transport of sand or the damage of gravel and rock increases. In general, a value of 0.035 is used for the critical Shields parameter in hydraulics when making a design. Although there is no guarantee that no transport or damage will occur, this value is generally accepted if combined with a monitoring strategy, i.e. if the authority will provide additional rock if nearby critical conditions occurred. In this respect, reference is made to Section 2.3 on risk assessment.

Hoffmans (2012) provides a probability density function and the distribution function (or cumulative probability density function) of the transport parameter (ratio between average bottom shear stress and average critical bed shear stress). If the failure probability has been determined, it is possible to determine the distribution function of the transport parameter and subsequently the critical Shields parameter.

The Shields figure, see Figure 4.8, is not very user friendly and therefore Hoffmans provided another way of presenting the critical Shields relation (see Figure 4.10). On the vertical axis, one can find the hydraulic load as the product of r_0 and U_0, and the horizontal axis represents the strength, i.e. relative density and stone size. Figure 4.10 shows an upper limit for $\Psi_c = 0.06$ and a lower limit for $\Psi_c = 0.02$. Nearly all experimental results are within the two limits, which is also consistent with the results of Paintal (1971); see Hoffmans (2012) for more information.

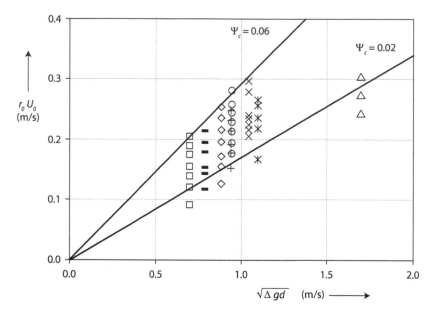

Figure 4.10 Shields diagram in the form of the relation between load and strength.

In the US, the critical bed shear stress is also written as a function of the grain size diameter d_{50} (HEC-18):

$$\tau_c = c\, d_{50} \tag{4.8}$$

with c (= 10^3N/m^3) which is a parameter with a constant value and a dimension. Combining Equations (4.6) and (4.7), the Shields equation can be written as follows:

$$\Psi_c = \frac{\tau_c}{(\rho_s - \rho)\, g d_{50}} \tag{4.9}$$

or with (4.8)

$$\Psi_c = \frac{c}{(\rho_s - \rho)\, g} = \frac{10^3}{(2650 - 1000)\cdot 9.81} = 0.06.$$

The critical Shields value for turbulent flows varies between 0.03 and 0.06. However, combining Equations (4.8) and (4.9) shows that using Equation (4.8), as performed in the US, implicitly means that a value of 0.06 is used for non-cohesive materials.

   Probably, Vanoni (1966) plays a role in this because he noted that the critical Shields parameter for the fully rough turbulent zone, i.e. $\Psi_c = 0.06$, corresponds to a low but measurable bed load. At values of 0.03 and even less, occasional movement of single grains may still occur. Conclusion: Equation (4.8) does not represent a practical criterion for the incipient motion of particles.

4.3.4 Critical flow velocity

For a uniform flow (logarithmic velocity profile) over a hydraulically rough bed, the critical mean velocity U_c is:

$$U_c = \frac{u_{*,c}C}{\sqrt{g}} \tag{4.10}$$

where the Chézy coefficient is given by:

$$C = \frac{\sqrt{g}}{\kappa} \ln\left(\frac{12R}{k_s}\right) \tag{4.11}$$

in which
   k_s = equivalent roughness of Nikuradse (m), hydraulically rough flow: $k_s = 3d_{90}$, hydraulically smooth flow: $k_s = 2d_{50}$
   R = hydraulic radius (m)
   $\kappa = 0.4$, constant of von Kármán (–)

If the width B of the flow is large compared to the flow depth h, U_c can be rewritten as follows:

$$U_c = 2.5\sqrt{\Psi_c \Delta g d} \ln\frac{12h}{k_s} \tag{4.12}$$

The equivalent roughness of a plane bed is usually related to the larger particles of the bed (d_{65}, d_{84} or d_{90}). The influences of the gradation, the shape of the particles and the flow conditions are generally disregarded. In systematic research (Delft Hydraulics, 1972) on scouring in which relatively uniform sediments were used, the equivalent roughness was assumed to be equal to the median particle diameter. However, in the literature, several expressions for k_s can be found (e.g. van Rijn, 1982).
   Van Rijn (1984) relates the critical Shields parameter to the nondimensional sedimentological diameter D_*:

$$D_* = d\left(\frac{\Delta g}{v^2}\right)^{1/3} \text{ with } \qquad v = \frac{40 \times 10^{-6}}{20 + \theta} \tag{4.13}$$

in which
   v = kinematic viscosity (in m^2/s)
   θ = temperature (in °C)

Empirical relations for Ψ_c, as proposed by van Rijn (1984), are presented in Table 4.4.
   According to van Rijn (1982), the equivalent roughness of a plane bed varies from 1 to 10 times d_{90} of the bed material. These rather large values indicate that a completely plane bed does not exist. For conditions with active sediment transport, the relatively large scatter of the equivalent roughness is probably caused by the initial unevenness (initial bed forms).

Table 4.4 Empirical relations for Ψ_c (van Rijn, 1984)

Ψ_c as a function of D_* (Equation 4.13)		
$\Psi_c = 0.24D_*^{-1}$	for	$D_* \leq 4$
$\Psi_c = 0.14D_*^{-0.64}$	for	$4 < D_* \leq 10$
$\Psi_c = 0.04D_*^{-0.10}$	for	$10 < D_* \leq 20$
$\Psi_c = 0.013D_*^{0.29}$	for	$20 < D_* \leq 150$
$\Psi_c = 0.055$	for	$D_* > 150$

Note: Not only sand and gravel are granular materials but also silt. It has a size somewhere between 0.002 and 0.063 mm, and its mineral origin is quartz and feldspar. Practically, the critical flow velocity for silt can be taken as 0 m/s ($U_c=0$) since its size is small and silt is non-cohesive.

4.3.5 Rock

Scour of rock may occur downstream of dam spillways, as a result of the impact of high-velocity jets and the resulting pressure fluctuations on the water–rock interface. The scour process at plunge pool beds is a result of the interactions of three phases involved: water, rock and air. Many empirical engineering methods are available to estimate scour formation downstream of plunging jets. However, some are based on physical backgrounds (see also Section 5.4, 'Plunging jets'). Besides these analytical relations, there are also mathematical models that predict rock scour in greater detail.

Rock scour can occur in four modes (NCHRP, 2011):

1. Dissolution of soluble rocks;
2. Cavitation;
3. Quarrying and plucking of durable, jointed rock, and
4. Abrasion and grain-scale plucking of degradable rock.

An example of rock scour is presented in Figure 4.11.

Rock scour by wall jets downstream of dams can compromise dam safety. The potential for such scour and its extent can be quantified by making use of the Erodibility Index Method (Annandale, 1995, 2006). This method is based on a relation between the stream power of flowing water and a geo-mechanical index representing the erosion resistance of any earth material, ranging from non-cohesive and cohesive soils to rock formations.

The Comprehensive Scour Model (CSM) was first proposed by Bollaert and Schleiss (2003).

It has the advantage of considering the physical phenomena involved in the scour of the rock impacted by plunging water jets. The model was developed as a result of experiments with plunging jets of near-prototype velocities, leaving an impact on closed-end and open-end fissures at the pool bottom. As such, the model reproduces the characteristics of the pressure signals of prototype jets, thus minimising scale effects.

Figure 4.11 **Scour in rock; sandstone (HEC 18).**

4.4 Cohesive Sediments

4.4.1 Introduction

This section gives a guide to the determination of the critical shear stress or critical flow velocity of cohesive soil materials. Tables with critical values are presented as well as formulas to estimate the critical values. In both cases, the different influences, such as plasticity index, water content and cohesion, are mentioned, because this make cohesive soils differ from non-cohesive soils (see also the previous Sections).

For cohesive sediments, relatively large forces are necessary to break the aggregates within the bed and relatively small forces are necessary to transport the material. Experiments by Mirtskhoulava (1988, 1991) have shown that clay soils with a natural structure in a water-saturated state are scoured in several stages.

In the initial stage, loosened particles and aggregates separate and those with weakened bonds are washed away. This process leads to the development of a rougher surface. Higher pulsating drag and lift forces increase the vibration and dynamic action on the protruding aggregates. As a result, the bonds between aggregates are gradually destroyed until the aggregate is instantaneously torn out of the surface and carried away by the flow.

The above mentioned scour process is influenced by the following parameters: cohesion, Cation Exchange Capacity (CEC), salinity, Sodium Adsorption Ratio (SAR), pH level of pore water, temperature, sand, organic content and porosity (e.g. Winterwerp, 1989).

In order to clarify qualitatively the erosion rate of a cohesive bed which is determined by the mutual effects of the sediment and pore water properties, some additional information is presented on parameters effecting the erosion rate. A parameter describing the properties is the SAR, which is indicative of the processes in the transition layer.

In general, the critical bed shear stress will increase with decreasing SAR and the critical bed shear stress will increase when salt is added to the pore water. An increase in temperature will decrease the strength of the bed. In addition to commonly determined parameters such as granulometry and mineralogy, the specific surface of the sediment and the CEC are also important. With a larger specific surface, the van der Waals forces become larger and the sediment becomes more cohesive. An increase in CEC with low SAR will also result in an increase in cohesion: the critical bed shear stress will increase and the erosion rate will decrease with increasing CEC.

In general, an increase in the organic content will increase the cohesiveness of the sediments, resulting in a larger critical bed shear stress and a lower erosion rate. However, this effect is known only qualitatively. No quantitative information is found in the literature. Another natural aspect is biological activity, such as bioturbation. The effect of sand on the strength of a cohesive bed seems to be dependent upon the value of the SAR: at low SAR, the strength of the bed will decrease with the increasing sand content, whereas the reverse trend will occur at high SAR.

In Sections 4.4.2 and 4.4.3, the critical shear stress and the critical flow velocity are addressed, respectively. In Section 4.4.4, empirical shear stress formulas are presented. The erosion rate can be estimated with a formula presented in Section 4.4.5. The stability of peat is discussed in Section 4.4.6.

4.4.2 Critical shear stress

Usually the incipient motion of sand is determined by horizontal shear stresses or horizontal forces. If the bed shear stress exceeds its critical value, then particles move in the streamwise direction. The Shields approach cannot be applied for cohesive materials, such as clay, peat and grass. Therefore, the stability of these soils is calculated by a turbulence approach which can easily be deduced to the original formula of Brahms (1767), according to Deltares (2015). The stability criterion reads (Hoffmans, 2012):

$$\frac{F_{lift}}{F_{down}} \leq \frac{12.6\rho\left(r_0 U_0\right)^2}{\left(\gamma_{wet}-\gamma\right)D+c} \text{ with } \gamma_{wet} = \left(1-n\right)\rho_s g + n\rho g \qquad (4.14)$$

in which
   c = cohesion (N/m^2)
   D = thickness of cohesive layer (m)
   F_{lift} = lift force (N)
   F_{down} = downward force (N)
   γ = weight of water per unit volume (kN/m^3)
   γ_{wet} = wet weight of soil per unit volume (kN/m^3)
   n = porosity (–)
   ρ_0 = relative turbulence intensity (-)
   U_0 = actual depth-averaged flow velocity (m/s)
   ρ = density of water (kg/m^3)
   ρ_s = material density (kg/m^3)

The soil is in equilibrium if $F_{lift}=F_{down}$ or in terms of shear forces if $F_{shear}=F_c$ or in terms of shear stresses if $\tau_0=\tau_c$ with:

$$\tau_0 = 0.7 \cdot \rho \cdot (r_0 U_0)^2 \tag{4.15}$$

and

$$\tau_c = 0.7 \cdot \rho \cdot (r_0 U_{0,c})^2 \tag{4.16}$$

in which
$U_{0,c}$ = critical depth-averaged flow velocity (m/s)

Instead of Equation (4.16), also the following formula for τ_c can be used:

$$\tau_c = \frac{1}{2} \cdot c_f \cdot \rho \cdot U_{b,c}^2 \tag{4.17}$$

in which
$U_{b,c}$ = water velocity just above the bed due to the return current (m/s)
c_f = resistance coefficient (–)
c_f = 0.010 (range 0.005–0.020)

The resistance coefficient c_f depends on the return current. Delft Hydraulics (1988) provides accurate equations for a final design. Equation (4.14) can be rewritten for $U_0 = U_{0,c}$ into:

$$\tau_c = \psi_c \left[(\rho_s - \rho) g D + c \right] \tag{4.18}$$

Equation (4.18) enables to compute critical values for bed materials. Section 4.5 presents examples. These examples show that the critical shear stress or critical flow velocity strongly depends on cohesion and turbulence intensity. Although the triaxial test gives information about the cohesion (c), and the angle of internal friction (ϕ), the Direct Simple Shear test (DSS test) is recommended for weak soils (clay, peat). A guideline is available for the DSS-test, viz. ASTM-D6528-17 (2017).

4.4.3 Critical flow velocity

In general, no readily applicable design equations for the depth of scour holes are available for cohesive sediments. The literature relates most equations to one or two particular parameters influencing the erosion of cohesive sediments. Moreover, they are often related to a specific sediment type. In many scour predictions, a critical velocity is applied, for example, the Breusers method.

For a first estimate, the following values may be used. For fairly compacted clay with a void ratio (i.e. the ratio between the volume of voids and the volume of the mineral part of soil) of 0.5, the critical depth-averaged velocity is about $U_c \approx 0.8$ m/s, whereas for stiff clay (a void ratio of 0.25) U_c is about 1.5 m/s. More values are presented in Table 4.5 based on expert judgement (CUR, 1996). The influence of the water depth is not mentioned in this table. For an increasing water depth, the critical flow velocity increases (see also Equation 4.19).

Table 4.5 Critical depth-averaged flow velocities for cohesive sediments (rough estimates)

Type of soil	U_c (m/s)
Loamy sand, light loamy clay with low compactness	0.4
Heavy loamy clay with low density	0.5
Low-density clay	0.6
Light loamy clay with medium compactness	0.8
Heavy loamy clay with medium density	1.0
Clay of medium density	1.3
Light loamy clay (dense)	1.2
Heavy loamy clay (dense)	1.5
Hard clay	1.9
Peat (no cohesion)	1
Peat (with cohesion)	1–2
Grass (poor quality)	1–3
Grass (good quality)	1.5–4
Grass (very good quality)	2–5

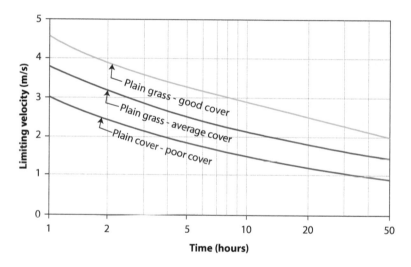

Figure 4.12 Critical flow velocity as a function of time (Hewlett et al., 1987).

The strength of grass can also be derived from the CIRIA diagram (Hewlett et al., 1987) in Figure 4.12. It describes the admissible depth-averaged flow velocity for the start of erosion of three types of grass covers as a function of time. It should be noted that the diagram underestimates the strength of the average grass cover because (see also the CIRIA report on the use of vegetation in civil engineering (2007)):

- The tested grass covers were relatively young; about 1 year (grass reaches its maximum strength after 4 years).
- Compared to the original measurements, the curves are conservative.

At transitions, the load on a rough bottom is substantially higher if water flows from a smooth bottom (stone pitching) to a rough bottom (grass cover), in particular close to the transition. Experimental research by Nezu (1977, 1993, 1994) showed that the bed shear stress at the transition is about twice as high. Hence, if the bed is not protected, erosion or scour will occur. Also the steepness of a slope will increase (see Section 3.4).

Within the Room-for-the-River programme in the Netherlands, grass covers have been applied in bypasses. The grass cover at transitions will be loaded extra. Hoffmans et al. (2014) discuss this aspect in detail. Based on extensive tests by Mirtskhoulava (1988, 1991), a simplified generally applicable expression for the critical depth-averaged flow velocity for cohesive sediments is:

$$U_c = \log\left(\frac{8.8h}{d_a}\right)\sqrt{\frac{0.4}{\rho}\left((\rho_s - \rho)gd_a + 0.6C_f\right)} \tag{4.19}$$

in which
    $C_f = 0.035C_o$ N/m^2, fatigue rupture strength of clay
    C_o = cohesion at saturation water content (Table 4.6), N/m^2
    d_a = size of detaching aggregates, $d_a = 0.004$ m
    h = flow depth (m)
    U_c = critical mean velocity for cohesive sediments (m/s)

Table 4.6 C_o in kPa and ϕ' in degrees (given between brackets); rough estimates (Mirtskhoulava, 1988)

Type of soil	Range of liquidity index	Soil property at void ratios						
		0.45	0.45	0.65	0.75	0.85	0.95	1.05
Loamy sand	0–0.25	14.7	10.8	7.85				
		(30)	(29)	(27)				
	0.25–0.75	12.7	8.83	5.88	2.94			
		(28)	(26)	(24)	(21)			
Loamy clay	0–0.25 (low plasticity)	46.1	36.3	30.4	24.5	21.6	18.6	
		(26)	(25)	(24)	(23)	(22)	(20)	
	0.25–0.5 (medium plasticity)	38.2	33.3	27.5	22.6	17.7	14.7	
		(24)	(23)	(22)	(21)	(19)	(17)	
	0.5–0.75 (high plasticity)			24.5	19.6	15.7	13.7	11.8
				(19)	(18)	(16)	(14)	(12)
Clay	0–0.25		79.4	66.8	53.0	46.1	40.2	35.3
			(21)	(20)	(19)	(18)	(16)	(14)
	0.25–0.5			55.9	49.0	42.2	36.3	31.4
				(18)	(17)	(16)	(14)	(11)
	0.5–0.75			44.1	40.2	35.3	32.4	28.4
				(15)	(14)	(12)	(10)	(7)

Note: void ratio $e = V_v/V_s$ with V_v is the volume of voids and V_s the volume of solids

Mirtskhoulava (1988, 1991) finds the cohesion at saturation water content to be the most significant features among the extensive complex of physio-mechanical and chemical properties of cohesive sediments. However, it does not suffice to rely on averaged properties. The inhomogeneity of the river bed must also be considered. The erosion characteristics of cohesive sediments remain poorly understood. For specific sediments at a given location, quantitative information relating to the erosion parameters is available, but for most situations, the designer has to perform an erosion test.

The parameters C_0 and ϕ are the most essential for determining the critical flow velocity of clay and peat (Figure 4.13). The most common test for determining the shear strength of soils is the direct simple shear (DSS) test. In the past also, the triaxial shear test was used for the determination of C_0 and ϕ but nowadays this test is only used for the determination of ϕ of loosely packed materials, e.g. sand.

In the DSS test (see Figure 4.14), a sample of soil is placed in a rectangular box, the top half of which is free to slide over the bottom half. The lid of the box is free to move vertically, and a normal stress σ_N is applied to the lid. A horizontal shearing stress τ is applied to the top half of the box, gradually increasing in strength until the soil begins to shear. The DSS test can be used to determine the strength of weak material. The US standard ASTM-D6528-17 (2017) gives directions on how to carry out the test.

In the triaxial shear test, a cylindrical soil sample is encased in a rubber membrane with rigid caps on top and bottom. The sample is placed in a closed chamber and subjected to a confined pressure on all sides. An axial stress is applied to the ends of the cylinder. The axial stress is either increased or the confining stress decreased until the sample fails in shear along a diagonal plane or number of planes.

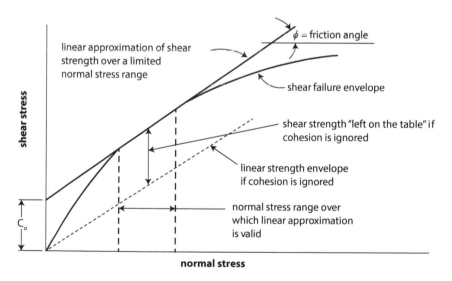

Figure 4.13 Exaggerated schematic of true curvilinear shear strength envelope, linear interpretation over a selected normal stress range, and the penalty for ignoring cohesion.

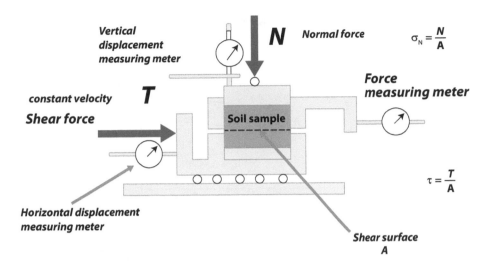

Figure 4.14 Principles of the direct simple shear test.

The shear strength is a function of confining stress and the direct shear, and the tests are used to find values for the shear strength parameters c and ϕ (see Figure 4.13). These parameters represent the coefficients of a straight line plotted through the results of similar tests on the same soil, with the only variable being changes in confining stress.

The tangent of ϕ is the coefficient of friction and represents the frictional component of soil strength, and c is the value of the intercept of the line, representing strength with no confining pressure (no friction). Under certain conditions, soils with significant fines content (especially clay) exhibit a significant c intercept and this is the source of their label as 'cohesive' soils.

4.4.4 Empirical shear stress formulas

Winterwerp et al. (2012) present a simple equation to estimate the value of τ_c:

$$\tau_c = 0.7 PI^{0.2} \quad \text{for} \quad PI > 7\% \tag{4.20}$$

where PI = plasticity index. The constant 0.7 (with the dimension of N/m^2) is an average in the range of 0.35–1.4. An average value for τ_c is 1.1 N/m^2. Smerdon and Beasley (1959) presented a formula for $PI < 7\%$ (in Winterwerp et al., 2012):

$$\tau_c = 0.163 PI^{0.84} \quad \text{for} \quad PI < 7\% \tag{4.21}$$

Both criteria are shown in Figure 4.15. In Section 4.5.6, an example is presented comparing the Winterwerp Equation (4.20) with the Mirtskhoulava Equation (4.19).

Note: for $PI < 7\%$, soil samples are low to non-cohesive.

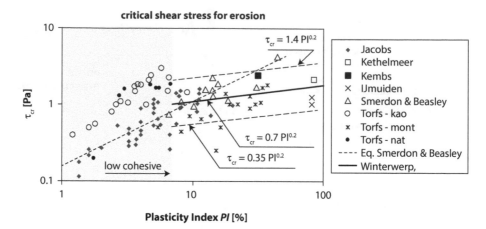

Figure 4.15 Critical shear strength for erosion as a function of the Plasticity Index (Winterwerp et al., 2012).

The value of *PI* can be estimated with (HEC-18) and Figure 4.16:

$$PI = 0.73(LL - 20) \tag{4.22}$$

in which
LL = liquid limit (–)

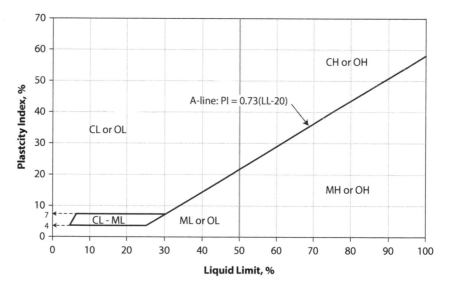

Figure 4.16 Plasticity chart for fine-graded soils (HEC-18)
Note: CH = clay with LL>50, CL = clay with LL<50
MH = silt with LL>50, ML = silt with LL<50
OH = organic material with LL>50, OL = organic material with LL<50

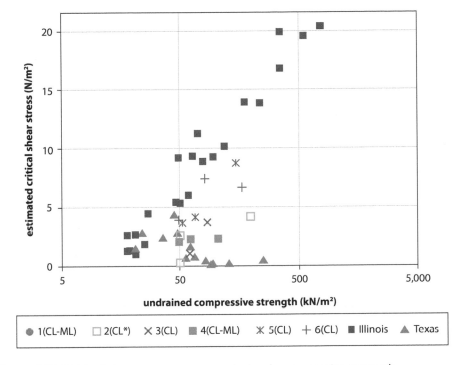

Figure 4.17 Critical shear stress versus undrained compressive strength
Note: CL = clay with LL<50, ML = silt with LL<50

Erosion resistant cohesive soils can be found above the lines in Figure 4.16. Recently, FHWA (2015) presented a design graph on the basis of tests (see Figure 4.17). The equation in the graph reads:

$$\tau_c = \alpha_c \left(\frac{w}{F}\right)^{2.0} PI^{1.3} f_u^{0.4}$$
(4.23)

where
  w = water content (–)
  F = fraction of fines of the soil smaller than 0.075 mm (–)
  f_u = undrained compressive strength (N/m^2)
  α_c = conversion constant, $\alpha_c = 0.07$

The coefficient of determination R^2 of the equation is 0.73. The figure legend provides the soil index number with an indicator of the soil classification CL (clay with LL<50) or CL-ML (clay/silt mixture with LL<50).

In Equation (4.23) and Figure 4.17, the undrained shear strength f_u is used. The undrained shear strength f_u can be determined with the DSS test (Figure 4.14). For practical purposes, the undrained shear strength f_u can be assumed to be equal to the cohesion C_0 that can be determined with a triaxial test. Although in reality f_u can be

Table 4.7 Soil strength of cohesive sediments (Sowers, 1970) (CUR 162)

Term	Undrained shear stress	Field test
Very soft	< 12.5 kPa	Squeezes between fingers when fist is closed
Soft	12.5–25 kPa	Easy moulded by fingers
Firm	25–50 kPa	Moulded by strong pressure of fingers
Stiff	50–75 kPa	Dented by strong pressure of fingers
Very stiff	75–100 kPa	Dented only slightly by finger pressure
Hard	>100 kPa	Dented only slightly by pencil point

smaller or larger than C_0 for practical engineering the undrained strength is assumed equal to the cohesion.

4.4.5 Erosion rate

For cohesive soils, a general erosion-rate formula can be used (Osman & Thorne, 1988):

$$\frac{dz}{dt} = \frac{R}{\rho_b \cdot g} \cdot \tau_c \left(\frac{\tau_0}{\tau_c} - 1 \right) \text{ and } R = 0.364 \cdot \tau_c \cdot e^{-1.3 \cdot \tau_c} \tag{4.24}$$

in which g=acceleration of gravity (m/s²), R=erosion parameter (kg/(m·s³)), t=time (s) and ρ_b=density of the bed material (kg/m³).

The erosion parameter R is valid for critical shear stresses (τ_c) larger than 0.6 Pa.

Note that the flow velocities in the scour hole decrease with time, and subsequently, the erosion rate will also decrease with time. In the equilibrium phase $\tau_0 = \tau_c$ and thus $dz/dt = 0$.

More sophisticated equations for the erosion rate of cohesive soils requiring more values of the input parameters can be found in the study by Winterwerp et al. (2012). The values of the bottom shear stress τ_0 and the critical bottom shear stress τ_b can be computed with Equations (4.15) and (4.16) or (4.17) respectively. Instead using Equations (4.16) or (4.17) the average value of τ_c follows also from Equations (4.20) and (4.21) (Winterwerp et al., 2012). If the bed consists of layers with different characteristics, the total scour should be computed for each layer separately with appropriate values for ρ and τ_c for that particular layer.

Karelle et al. (2016) used a simple formula based on Briaud (2008) for the erosion rate in a specific case of a jet impinging on a slope:

$$\frac{dz}{dt} = \alpha U^\beta \tag{4.25}$$

in which U is flow velocity (m/s). The erosion rate dz/dt is given in mm/hr. According to Briaud, the value of α varies between 1.2 (clay) and 11.8 (sand) and the value of β between 5.6 (sand) and 2.9 (clay). The values of α and β depend on the type of soil, see also Figure 4.7 where the erosion rate depends on the type of material.

4.4.6 Peat

Peat soil occurs in many areas and generally originates from plant and animal remains. It is considered partly as decomposed biomass (Adnan & Wijeyesekera, 2007).
Due to this composition, the structure of this soil differs from inorganic soils like clay,
sand and gravel. Peat has a high compressibility, low shear strength, high moisture
content and low bearing capacity (Adnan et al., 2007).

Various researchers have studied the shear strength and the reported results indicate important differences in behaviour of inorganic soils both qualitatively and quan-
titatively. Following Den Haan (1997), both undrained shear strength and effective
strength parameters of many peats increase with increasing water content or decreasing unit weight. This result seemingly supports the intuitive behaviour that can be
attributed to the fibre effects and the fact that the fibre content, in general, increases
with the increasing water content and decreasing unit weight.

Hoffmans (2012) derived the stability criterion presented in Equation (4.14):

$$\frac{F_{lift}}{F_{down}} \leq \frac{12.6\rho\left(r_0 U_0\right)^2}{\left(\gamma_{wet} - \gamma\right)D + c} \text{ with } \gamma_{wet} = \left(1 - n\right)\rho_s g + n\rho g \qquad (4.14)$$

Default values for the parameters for peat are as follows:
$c = 0$–2.5 N/m^2
$D = 0.2$ m
$\gamma = 10$ kN/m^3
$\gamma_{wet} = 10.5$–12 kN/m^3
$n = 0.4$ (–)

In Section 4.4.2, already the critical shear stress is given (Equation 4.18). This equation can easily be rewritten into a formula for the critical flow velocity (see Hoffmans,
2012):

$$U_c = r_0^{-1} \sqrt{\frac{\left(\gamma_{wet} - \gamma\right)D + c}{12.6\rho}} \qquad (4.26)$$

The stability of peat against scour can also be computed with the empirical Equations
(4.20)/(4.21) or (4.23). As shown already, the critical flow velocity depends strongly on
cohesion and turbulence intensity. However, the formula gives only the first estimate
since validation is very limited. Based on published values for the cohesion of peat
(Kirk, 2001), the low and high representative values are 2.5 and 17.5 kN/m^2, respectively. The average representative value measures 10 kN/m^2. The strength of peat can
be determined with the DSS test.

Examples of peat scour are shown in Figures 4.18 and 4.19.

Huismans et al. (2016) discussed the scour hole development in river beds with mixed
sand–clay–peat stratigraphy. This phenomenon plays particularly in river areas. The
scour can be predicted with the equations of Hoffmans (2012); see Section 4.4.2.

Figure 4.18 **Floating peat islands in Gouwekanaal, near Julianasluis.**

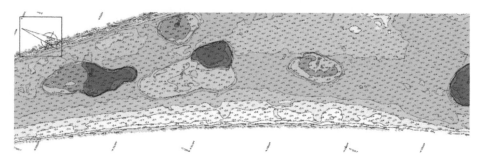

Figure 4.19 Local scour in Gouwe kanaal due to navigation.

4.5 Examples

4.5.1 Introduction

A number of examples will be treated in this section. First, in Section 4.5.2, the relative turbulence intensity is computed at a bridge pier and near a groyne. For a bridge pier, it is shown that the types of bridge pier and the scour depth are important for the relative turbulence intensity. For a groyne, the magnitude of the turbulence coefficient C_k is important, i.e. the steepness of the upper slope. Section 4.5.3 deals with the determination of the critical flow velocity of peat. The strong effect of the relative turbulence intensity and cohesion is shown. In Section 4.5.4, the critical flow velocity and the critical bed shear stress are computed in an open channel for a bed consisting of sand dunes. The roughness of the dunes plays a role. Section 4.5.5 deals with the Mirtskhoulava formula for cohesive materials. The critical flow velocity is computed for a clay using various sources of information. Section 4.5.6 also addresses initiation of motion of clay. The computation of the critical shear stress is shown using both the method of Winterwerp and the Mirtskhoulava formula.

4.5.2 Turbulence at bridge piers and groynes

Two examples show how to determine the relative turbulence intensity r_0 in case of a scour hole at a bridge pier and near groynes.

4.5.2.1 Bridge Piers

Assuming that the equilibrium scour depth equals the upstream water depth, thus $y_{m,e}=h_0$, the depth-averaged flow velocity in the scour hole reads: $U=0.5\ U_0$. In case of a relatively wide pier, the depth-averaged relative turbulence intensity follows with Equations (4.3)–(4.5) and $C_k=0.045$:

$$r_{0,m}=\frac{\sqrt{k_{gem}}}{U}=\frac{\sqrt{\tfrac{1}{2}C_kU_0^2}}{\tfrac{1}{2}U_0}=2\sqrt{\tfrac{1}{2}C_k}=2\sqrt{\tfrac{1}{2}\cdot0.045}\approx0.30$$

Assuming that the water depth is about 10 times the scour depth, or neglecting the scour depth, the depth-averaged flow velocity in the scour hole is $U\approx U_0$. This holds for slender bridge piers where the scour depth is small compared to the water depth. The result is:

$$r_{0,m}=\frac{\sqrt{k_{gem}}}{U}=\frac{\sqrt{\tfrac{1}{2}C_kU_0^2}}{U_0}=\sqrt{\tfrac{1}{2}\cdot0.045}\approx0.15$$

This means that the depth-averaged relative turbulence intensity in the scour hole decreases if the bridge pier becomes more slender. Summarising: in the scour hole, $r_{0,m}$ varies between 0.15 (slender piers; lower limit) and 0.30 (wide piers; upper limit).

Now consider a circular bridge pier in a river with a water depth of 10 m during a flood. The width of the pier is 4 m. We estimate the equilibrium scour depth to be 6 m (1.5 x 4 m). This means the depth-averaged flow velocity in the scour hole reads $U=10/(10+6)U_0=U_0/1.6$ and subsequently,

$$r_{0,m}=\frac{\sqrt{k_{gem}}}{U}=\frac{1.6\sqrt{\tfrac{1}{2}C_kU_0^2}}{U_0}=1.6\sqrt{\tfrac{1}{2}\cdot0.045}\approx0.24$$

4.5.2.2 Groynes

For the condition that the equilibrium depth equals half the upstream water depth, thus $y_{m,e}=\tfrac{1}{2}h_0$, the depth-averaged flow velocity in the scour hole is $U=2/3U_0$. With $C_k=0.045$, this leads to:

$$r_{0,m}=\frac{\sqrt{k_{gem}}}{U}=\frac{\sqrt{\tfrac{1}{2}C_kU_0^2}}{\tfrac{2}{3}U_0}=1.5\sqrt{\tfrac{1}{2}C_k}=1.5\sqrt{\tfrac{1}{2}\cdot0.045}\approx0.23$$

Note: $C_k = 0.045$ is valid for steep upstream slopes. For less steep slopes, a value of $C_k = 0.03$ should be used, resulting in:

$$r_{0,m} = \frac{\sqrt{k_{gem}}}{U} = \frac{\sqrt{\frac{1}{2} C_k U_0^2}}{\frac{2}{3} U_0} = 1.5\sqrt{\frac{1}{2} C_k} = 1.5\sqrt{\frac{1}{2} \cdot 0.03} \approx 0.18$$

Example: During a flood, the water depth is 10 m. The depth-averaged flow velocity in the river and at the location of the maximum scour depth are 2.0 and 1.2 m/s, respectively (thus, $U = 3/5 U_0$). For live-bed scour, the relative turbulence intensity $r_{0,m}$ with $C_k = 0.03$ is:

$$r_{0,m} = \frac{\sqrt{k_{gem}}}{U} = \frac{\frac{5}{3}\sqrt{\frac{1}{2} C_k U_0^2}}{U_0} = \frac{5}{3}\sqrt{\frac{1}{2} \cdot 0.03} \approx 0.2$$

In addition to the above formulas to compute $r_{0,m}$, a more general formulation of the turbulence levels and damping of the turbulence in the flow direction is required for application downstream of structures. A user-friendly general formulation is now available, based on geometrical characteristics and the continuity and transport equations for a sill only (see Equation 6.11). The formula can also be applied for the decrease of turbulence downstream of a hydraulic jump.

4.5.3 Critical flow velocity of peat

This example shows the strong effect of turbulence intensity and cohesion on the critical flow velocity. Moreover, it is shown that the same result would be obtained if Shields were used. Suppose we have peat in thin layers in a canal bed. The cohesion can be neglected, and thus $C_0 = 0$ kPa. Furthermore, $D_{peat} = 0.2$ m, $r_0 = 0.1$ (normal turbulence) and $\gamma_{wet} = 10.5$ kN/m³. The critical flow velocity can be derived from Equation (4.14) for $F_{lift} = F_{down}$ and $U_0 = U_c$ and reads:

$$U_c = \frac{1}{r_0}\sqrt{\frac{(\gamma_{wet} - \gamma)D + C_0}{12.6\rho}} = \frac{1}{0.1}\sqrt{\frac{(10500 - 10000) \cdot 0.2 + 0}{12.6 \cdot 1000}} \approx 0.9 \text{ m/s}$$

If $C_0 = 0.5$ kPa, the result is:

$$U_c = \frac{1}{r_0}\sqrt{\frac{(\gamma_{wet} - \gamma)D + C_0}{12.6\rho}} = \frac{1}{0.1}\sqrt{\frac{(10500 - 10000) \cdot 0.2 + 500}{12.6 \cdot 1000}} \approx 2.2 \text{ m/s}$$

If we increase the turbulence intensity to $r_0 = 0.2$, the result is:

$$U_c = \frac{1}{r_0}\sqrt{\frac{(\gamma_{wet} - \gamma)D + C_0}{12.6\rho}} = \frac{1}{0,2}\sqrt{\frac{(10500 - 10000) \cdot 0.2 + 500}{12.6 \cdot 1000}} \approx 1.1 \text{ m/s}$$

We can compare this with sand with $d = 0.2$ mm. Assume $C_0 = 0$ kPa, $r_0 = 0.1$ and $\gamma_{wet} = 20$ kN/m³, and then the critical flow velocity is (sand under water is always saturated):

$$U_c = \frac{1}{r_0} \sqrt{\frac{(\gamma_{wet} - \gamma)d + C_0}{12.6\rho}} = \frac{1}{0.1} \sqrt{\frac{(20000 - 10000) \cdot 0.0002 + 0}{12.6 \cdot 1000}} \approx 0.126 \text{ m/s}$$

This critical value complies with experimental values; see Equation (4.7) and (4.27) with $r_0 = 1.2g^{0.5}/C$ (Hoffmans, 2012):

$$U_c = \frac{1.2}{r_0} \sqrt{\Delta g d \Psi_c} = \frac{1.2}{0.1} \sqrt{1.65 \cdot 9.81 \cdot 0.0002 \cdot 0.033} \approx 0.12 \text{ m/s}$$

This results in:

$$\frac{1}{r_0} \sqrt{\frac{(\gamma_{wet} - \gamma)d + C_0}{12.6\rho}} = \frac{1.2}{r_0} \sqrt{\Delta g d \Psi_c}$$

or

$$\Psi_c = \frac{1-n}{18} = \frac{1-0.4}{18} \approx 0.033$$

Note that a critical Shields value of 0.033 is conservative.

In conclusion, the critical flow velocity strongly depends on the cohesion and the turbulence intensity. The DSS test is recommended to determine the cohesion and the angle of repose or internal friction for weak soils (clay, peat).

4.5.4 Critical mean flow velocity and critical bed shear stress in an open channel with sand dunes

A wide, open channel has a flow depth of $h = 5$ m. The bed is covered with sand dunes, and the bed material characteristics are $d_{50} = 300$ μm, $d_{90} = 500$ μm and $\rho_s = 2650$ kg/m³. The water temperature is 20°C ($v = 10^{-6}$ m²/s and $\rho = 1000$ kg/m³).

First, the sedimentological diameter is:

$$D_* = d_{50} \times (D \times g/v^2)^{1/3} = 300 \times 10^{-6} \times \left[1.65 \times 9.8/(10^{-6})^2 \right]^{1/3} = 7.6$$

The critical Shields parameter follows with:

$$\Psi_c = 0.14 \times D_*^{-0.64} = 0.14 \times 7.6^{-0.64} = 0.038 \ (4 < D_* < 10)$$

This gives a critical mean velocity:

$$\begin{aligned} U_c &= \sqrt{(\Psi_c \times \Delta \times g \times d_{50})} \big/ \kappa \times \ln[12 \times h/(3 \times d_{90})] \\ &= \sqrt{(0.038 \times 1.65 \times 9.8 \times 300 \times 10^{-6})} \big/ 0.4 \times \ln\left[12 \times 5 \big/ (3 \times 500 \times 10^{-6}) \right] \\ &= 0.36 \text{ m/s} \end{aligned}$$

$$C = \sqrt{g}/\kappa \times \ln[12 \times h/(3 \times d_{50})] = \sqrt{9.81}/0.4 \times \ln\left[12 \times 8/(3 \times 400 \times 10^{-6}) \right] = 88.4 \text{ m}^{1/2}/s$$

The critical bed shear-stress τ_c follows with:

$$\tau_c = \rho \times \left(u_{*,c}\right)^2 = \rho \times \Psi_c \times \Delta \times g \times d_{50}$$
$$= 1000 \times 0.038 \times 1.65 \times 9.8 \times 300 \times 10^{-6} = 0.184 \text{ Pa}$$

4.5.5 Critical depth-averaged flow velocity according to Mirtskhoulava (1988)

A channel has a constant discharge of $Q=12\,\text{m}^3/\text{s}$ and a flow depth of $h=1\,\text{m}$. The liquidity index (i.e. the difference between liquid limit and plastic limit, cf. ASTM, 1992) lies in the range of 0.25–0.50, and the void ratio is about 0.8. The channel bed consists of heavy homogeneous loamy clay of medium density. Other data are $\rho_s,=2650\,\text{kg/m}^3$, $\rho=1000\,\text{kg/m}^3$, $\theta=20°\text{C}$, $v=10^{-6}\,\text{m}^2/\text{s}$.

The first estimate for the critical mean flow velocity with Table 4.5 gives $U_c \approx 0.8$–$1.0\,\text{m/s}$.

Applying Table 4.6, the cohesion is about $C_o \approx 20\,\text{kPa}$.

However, the average cohesion of the cohesive sediment has been determined in a geotechnical laboratory and amounts to $C_o=15.5\,\text{kPa}$. Based on the work of Mirtskhoulava (1988, 1991), a simplified expression for the critical depth-averaged flow velocity for cohesive sediments is:

$$U_c = \log\left(\frac{8.8h}{d_a}\right)\sqrt{\frac{0.4}{\rho}\left((\rho_s - \rho)gd_a + 0.6C_f\right)} \tag{4.19}$$

Substituting the relevant values with $C_f=0.035C_o\,\text{N/m}^2$ and $d_a=0.004\,\text{m}$ gives:

$$U_c = \log(8.8 \times h/d_a)$$
$$\times \sqrt{\left[(0.4/\rho)\times\left((\rho_s - \rho)\times g \times d_a + 0.6 \times 0.035 \times C_o\right)\right]}$$
$$= \log(8.8 \times 1/0.004)$$
$$\times \sqrt{\left[(0.4/1000)\times\left((2650 - 1000)\times 9.8 \times 0.004 + 0.6 \times 0.035 \times 15500\right)\right]}$$
$$= 1.3 \text{ m/s}$$

4.5.6 Comparison critical strength of clay

This example compares two methods:

- Simplified method of Winterwerp et al. with Equation (4.20)
- Method of Mirtskhoulava with Equation (4.19).

Equation (4.19) results for $h=6.0\,\text{m}$, $\rho_s=2000\,\text{kg/m}^3$, $\rho=1000\,\text{kg/m}^3$, $g=9.8\,\text{m/s}^2$ and $C_0=20\,\text{kN/m}^2$ in

$$U_c = 1.75 \text{ m/s}$$

This value is comparable using Table 4.5, resulting in $U_c=1.5\,\text{m/s}$ for a heavy loamy clay.

Next, the Winterwerp method is applied and the critical shear stress will be translated into a critical flow velocity to make a comparison possible with the critical flow velocity of Mirtskhoulava. The Winterwerp equation for $PI > 7\%$ reads

$$\tau_c = 0.7 PI^{0.2} \tag{4.20}$$

The constant 0.7 (with the dimension of N/m^2) is the average in the range of 0.35–1.4. Winterwerp mentions for a range of PI values an average value for τ_c of 1.1 N/m^2.

Suppose $PI = 35\%$, and then Equation (4.20) results in $\tau_c = 1.4\,N/m^2$ (range 0.7–2.9 N/m^2 for constant 0.35 or 1.4)

A critical shear stress can also be estimated using Figure 4.17. Assuming an undrained compressive strength of $f_u = 20\,kN/m^2$ (f_u is almost equal to C_0), a critical shear stress can be estimated of $\tau_c = 1.4$–2.4 N/m^2 (or 0.03–0.05 lbf/ft^2) which is more or less in agreement with the value of 1.4 with Equation (4.20).

With Equation (4.16), we can translate the critical shear stress into a critical flow velocity using $r_0 = 0.1$, $\rho = 1000\,kg/m^3$ and $\tau_c = 1.4\,N/m^2$:

$$U_c = r_0^{-1} \sqrt{\frac{\tau_c}{0.7 \cdot \rho}} = (0.1)^{-1} \sqrt{\frac{1.4}{0.7 \cdot 1000}} = 0.45 \text{ m/s}$$

This critical flow velocity is 4 times smaller than computed with the Mirtskhoulava method. For critical shear stress values of about 20–25 N/m^2, critical flow velocities of 1.75 m/s can be computed. Perhaps the value of the bulk density is the reason of the difference between the two methods. Winterwerp et al. (2012) investigated soft soils with relative low bulk densities (see Figure 4.20) and deduced engineering tools for the critical bed shear stress as discussed. Mirtskhoulava examined the erodibility of different clay types with much higher bulk densities. Therefore, also the bulk density is a relevant erosion parameter.

Conclusion: derivation of a critical value for shear stress requires careful considerations. Small differences in parameter values may result in strong differences.

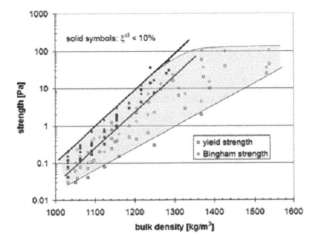

Figure 4.20 Measured values of strength for a variety of mud samples from the Netherlands (Winterwerp et al., 2012)

Chapter 5

Jets

5.1 Introduction

Jets can occur as a result of flow under, through or over hydraulic structures. In general, a jet lifts the sediment particles and transports those particles downstream of the impacted area. The jet impact area is transformed into an energy dissipator, and a scour hole is formed. This chapter discusses several forms of jets including plunging jets, submerged jets, horizontal and vertical jets, and two-dimensional (2D) and three-dimensional (3D) jets. Literature reviews of empirical relations are given by, for example, Whittaker and Schleiss (1984), Mason and Arumugam (1985), Breusers and Raudkivi (1991), and Hoffmans (2012).

Till date, however, there is no universal scour formula that is capable of handling the different flows downstream of hydraulic structures. The erosion process is complex and depends upon the interaction of hydraulic and morphological factors. Section 5.2 concerns the flow characteristics of jets and the surrounding mass. Section 5.3 pays attention to the scour time scale. Scour due to plunging jets is discussed in Section 5.4. Scour downstream of 2D and 3D culverts is treated in Sections 5.5 and 5.6, respectively. Section 5.7 addresses ship-induced flow and erosion. Section 5.8 deals with scour due to broken pipelines. Section 5.9 discusses some measures to control scour. Finally, some examples are discussed in Section 5.10.

5.2 Flow characteristics

5.2.1 Introduction

The scour associated with hydraulic structures may be caused by different types of jets, including plunging jets that impinge on an erodible bed and horizontal jets eroding bed material immediately downstream of structures. In addition, a distinction can be made between plane and circular jets and between submerged and unsubmerged jets. Numerous textbooks (e.g. Schlichting, 1951; Rajaratnam, 1976) discuss theoretical relations that predict the flow velocity in jets and the form of hydraulic jumps. In the next sections, some results of these studies are presented so far as they are relevant to scour.

5.2.2 Flow velocities

Generally, a mixing layer occurs between two streams that move at different speeds. Such a discontinuity in the flow is unstable and gives rise to a zone of turbulent mixing downstream of the point where the two streams meet first. The width of this mixing region increases in the downstream direction. The submerged jet flow can be divided into two distinct regions: the potential core and the fully developed flow region (Figure 5.1).

In the first part of the jet, the flow velocity equals the efflux velocity. In the fully developed part of the jet, the velocities decrease. In this part of a circular jet (3D), i.e. for $x > 6b_u$, the velocity distribution can be given by the internationally and generally accepted formula (e.g. Rajaratnam, 1976):

$$u = u_m e^{-72\eta^2} \quad \text{with} \quad \frac{u_m}{U_1} = 6.0 \frac{b_u}{x} \tag{5.1}$$

where
b_u = diameter of pipe or thickness of jet at $x = 0$ (m)
u = mean velocity in the x-direction (m/s)
u_m = maximum velocity of u at any x-section (m/s)
U_1 = efflux velocity at $x = 0$ (m/s)
x = longitudinal distance (m)
$\eta = z/x;$ z is the vertical distance from the axis of the jet at any section

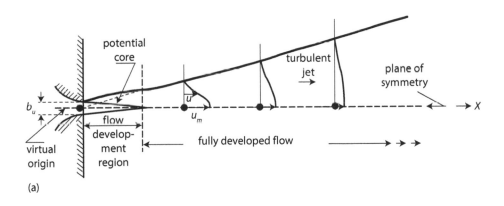

(a)

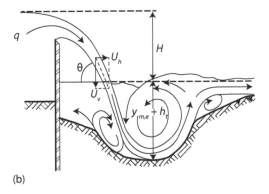

(b)

Figure 5.1 Characteristics of jet flow.

Table 5.1 Values of velocity coefficient (Przedwojski et al., 1995)

H(m)	1	2	3	4	5
c_v(—)	0.965	0.930	0.895	0.870	0.855

For plane jets (2D), the decrease in the maximum velocity is less and can be given by the internationally and generally accepted formula (for $x > 6b_u$):

$$u = u_m e^{-56\eta^2} \quad \text{with} \quad \frac{u_m}{U_1} = \sqrt{6.0 \frac{b_u}{x}}$$

(5.2)

Reference is made to Rajaratnam (1976) for the velocities in wall jets.

It is emphasised that the area with values of $x < 6b_u$ is called the flow development zone with the potential core (see Figure 5.1). In the potential core, the flow velocity is equal to the efflux velocity.

The decay of the turbulence downstream of the outflow opening is discussed in Section 4.2 (see also Equation (5.17) in Section 5.5.2). The relative turbulence can be computed for every distance behind the outflow opening by substituting x for L.

The sharp-crested weir is the simplest form of an overflow spillway. The shape of the flow nappe over a sharp-crested weir can be interpreted by using the projectile principle. Following this principle, the horizontal velocity component of the flow is constant and the only force acting on the nappe is gravity. The vertical component of the jet velocity U_v of a plunging jet at the entry point is assumed to be related to the difference in head H (Figure 5.1) (e.g. Przedwojski et al., 1995):

$$U_v = c_v \sqrt{2 g H}$$

(5.3)

which is the Torricelli law for the vertical velocity in a jet.

g = acceleration due to gravity, 9.8 m²/s (range 9.78–9.83 m/s²)

The angle θ of the jet entry follows from the geometrical description of the angle between the jet and the horizontal plane:

$$\theta = \arctan \left(U_v / U_h \right)$$

(5.4)

where U_h is the horizontal component of the jet velocity. The coefficient c_v is sometimes assumed to be unity. However, a smaller value, as presented in Table 5.1, is recommended to account for the energy loss.

5.2.3 Hydraulic jump

Downstream of hydropower turbines, for example, the flow above a bed protection generally consists of a supercritical part upstream, a hydraulic jump, and a subcritical part downstream. Usually, the bed protection consists of a concrete plate equipped with energy dissipaters to decrease the high turbulence intensity at the end of the bed protection. A hydraulic jump occurs when high velocities discharge into a zone of lower velocity. This transition from supercritical to subcritical flow is characterised by an abrupt rise in the water surface profile, violent turbulence and a large energy loss. Usually a hydraulic jump is associated with a downstream control (subcritical flow) and an upstream control (supercritical flow). If the Froude number in the jet is greater than 1.0, the tailwater depth can be given by the internationally accepted Bélanger equation (e.g. Schoklitsch, 1935):

$$h_t = \frac{1}{2} b_u \left(\sqrt{1 + 8\ Fr_1^2} - 1 \right) \tag{5.5}$$

where Fr_1 = Froude number in the jet, $U_1/(gb_u)^{1/2}$.

Although the length of the hydraulic jump is an important parameter, it is not possible to derive it from theoretical considerations. Values are in the range of $L_r/h_t = 4$–6. The hydraulic jump parameters depend upon the upstream Froude number, as can be seen in Figure 5.2.

When using a hydraulic jump as an energy dissipater in the design of a stilling basin, the jump position, the jump types and the tailwater conditions have to be considered. For scour protection purposes, a jump that occurs immediately ahead of the subcritical flow is an ideal case. However, in most practical problems, the tailwater fluctuates owing to changes in the discharge. When the tailwater is greater than the critical depth $h_c = (q^2/g)^{1/3}$, the jump will be forced upstream and may finally become a submerged jump. This is possibly the safest case in design because the position of the submerged jet is fixed. Unfortunately, such a design is not efficient since little energy will be dissipated.

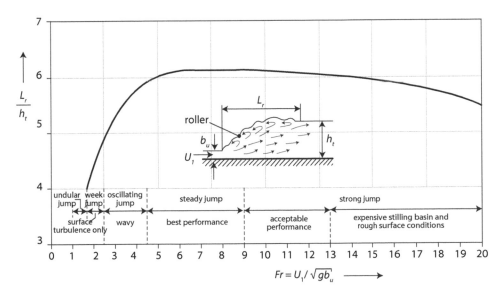

Figure 5.2 **Length of hydraulic jump (US Bureau of Reclamation, 1955).**

5.3 Time scale of jet scour

Farhoudi and Smith (1982, 1985) extensively studied scour below a spillway. A hydraulic jump was formed owing to a supercritical flow directed over a crest (Figure 5.3). The crest could be considered as a special kind of short-crested sill. The length of the apron was about equal to the length of the hydraulic jump. A plain apron without appurtenances was used. Three tailwater categories were tested: a submerged hydraulic jump ($h_t > d_1$, where d_1 is conjugate depth), a balanced hydraulic jump and a downstream-moved hydraulic jump ($h_t < d_1$).

The results were in general agreement with Breusers' work. The time development of the scour depth could be written as:

$$\frac{y_m}{d_0} = \left(\frac{t}{t_1}\right)^{\gamma}$$

(5.6)

where
   d_0 = characteristic length: $d_0 = 1/2\ h_d$ where h_d is the drop height (m)
   t = time (s)
   t_1 = characteristic time at which $y_m = d_0$ (s)
   y_m = maximum scour depth at t (m)
   γ = coefficient (–), $\gamma = 0.22$–0.23

Equation (5.6) has been calibrated and validated with experiments at scale model tests, but it has not been verified with prototype tests and thus is not generally applicable.

The value of γ in Equation (5.6) measured approximately $\gamma = 0.2$ and was scarcely affected by the tailwater conditions. The time scale between model and prototype was represented by:

$$n_t = n_{h_t}^2 n_{\Delta}^{1.4} n_{Fr_1}^{0.9} n_{(\alpha_F U_0 - U_c)}^{-3}$$

(5.7)

where
   U_c = critical mean velocity (m/s)
   U_0 = mean velocity (m/s), $U_0 = Q/A$, Q is discharge (m³/s), A is cross-sectional area (m²)
   α_F = turbulence coefficient (–), $\alpha_F = (1 + 3r_0)c_v$, where c_v is a velocity distribution coefficient (–), $c_v = 1.0$, r_0 is relative turbulence intensity (–)
   Δ = relative mass density (–)

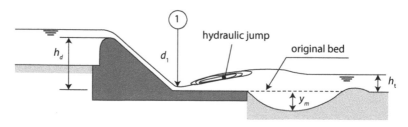

Figure 5.3 Definition sketch for model tests.

It should be noted that Equation (5.7) does not fulfil the Partheniades erosion law (see also Section 5.5.2).

Four sand sizes (0.15, 0.25, 0.52 and 0.85 mm), two bakelite sizes (0.25 and 0.52 mm) and three dam heights (0.1, 0.2 and 0.4 m) were employed. Scour profiles were found to be quite similar for all values of t. For a rock protective apron or for an apron with an end sill, the apron may be regarded as rough and, from Breusers' work, c_v may be taken as unity.

Directly downstream of a hydraulic jump, r_0 is about 0.6 and $r_0 = 0.3$ at a distance of 5 times the flow depth from the hydraulic jump. The time development of the maximum scour depth downstream of an outlet has been determined in different scale model investigations by Delft Hydraulics (1986a, 1986b, 1987, 1989). In all these investigations, structures with rectangular outlets, diverging wing walls and a bed protection of riprap were tested. The value of α was deduced from the measured scour depths and ranged from 2.5 to 3.5 (see also Sections 3.4.2 and 6.3). The scour holes due to circular outlets with a relatively high head difference are similar to the scour holes due to free falling jets.

5.4 Plunging jets

5.4.1 Introduction

The term 'plunging jets' refers to jets of water that impinge on the free surface due to discharge from an outlet above the free surface or overflow through an opening of a dam. The construction of large dams is invariably associated with the need to periodically release water downstream. This may be to generate electricity, to provide irrigation or to discharge flood waters which exceed the reservoir storage capacity. Owing to its frequent occurrence in engineering applications, the scour downstream of hydraulic structures constitutes an important field of research (Figure 5.4). To ensure the stability of the structure stilling basins, diversion works and sills without bed protection are built.

5.4.2 Calculation methods

Situations where fluid flows over completely submerged structures, such as flow over a sill with bed protection, are treated in Chapter 6. One of the earliest relations defining the scour depth due to a plunging jet on an unprotected bed was proposed by Schoklitsch (1932). Based on hundreds of flume experiments for the overflow type only, he arrived at the following empirical relation:

$$y_{m,e} + h_t = c_s q^{0.57} H^{0.2} d_{90}^{-0.32} \tag{5.8}$$

where
c_s = coefficient with a dimension, $c_s = 4.75 \ (m^{0.16} s^{0.57})$
d_{90} = particle diameter for which 90% of the mixture is smaller (m)
q = discharge per unit width (m²/s)
$y_{m,e}$ = equilibrium scour depth (m)

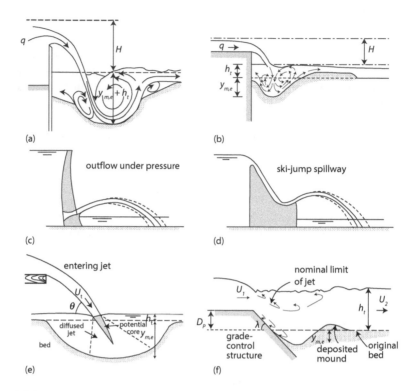

Figure 5.4 Jet types.

Both uniform and graded sediments were used to study the scour process down-stream of weirs. In each experiment, the tailwater depth was controlled to maintain the same value as the head above the weir (see Figure 5.4). Though the prediction accuracy of Schoklitsch's formula is high for flume experiments it is not recommended to use it for prototype conditions as discussed in Hoffmans (2012).

Bormann (Bormann & Julien, 1991) investigated local scour downstream of grade-control structures. The theoretical investigation was based on 2D jet diffusion and particle stability. According to Bormann and Julien (1991), the equilibrium scour depth in prototype conditions can be given by (see Figure 5.4; 2D-V jets):

$$y_{m,e} + D_p = \frac{K_b q^{0.6} U_1 \sin\theta}{(2\Delta g)^{0.8} d_{90}^{0.4}} \tag{5.9}$$

where
 D_p = drop height of grade-control structure (m)
 g = acceleration due to gravity, $g = 9.8\,\text{m/s}^2$
 K_b = coefficient (−)
 U_1 = jet velocity entering tailwater (m/s), $U_1 = \sqrt{2gH}$
 Δ = relative mass density (−)
 θ = jet angle near surface

The coefficient K_b is related to the jet angle θ and the angle of repose $\phi\,(= 25°)$. It reads:

$$K_b = c_d^2 \left(\frac{\sin\phi}{\sin(\phi+\theta)} \right)^{0.8} \quad \text{with } c_d = 1.8 \tag{5.10}$$

The scour relation thus derived has been experimentally calibrated with large-scale experiments. These experiments were conducted with unit discharges ranging from 0.3 to 2.5 m²/s and maximum scour depths reaching 1.4 m. The prediction accuracy of the formula is about 77% for a discrepancy ratio r between the predicted and measured value in the range of 0.67–1.5.

Hoffmans (2009) slightly modified a formula presented by Fahlbusch (1994) for the scour depth in the equilibrium stage for vertical 2D plunging jets. On the basis of a vertical balance of forces, the following equation was derived (see Figure 5.4):

$$y_{m,e} + h_0 = c_{2V} \sqrt{q U_1 \sin\theta / g} \tag{5.11}$$

$$c_{2V} = \begin{cases} \dfrac{20}{(D_{90*})^{1/3}} & \text{for } d_{90} \leq 1.25\,\text{cm} \\[2mm] 2.9 & \text{for } d_{90} \geq 1.25\,\text{cm} \end{cases}$$

$$D_{90*} = d_{90} \cdot \left(\frac{\Delta g}{v^2} \right)^{1/3}$$

where
c_{2V} = dimensionless parameter (-)
D_{90*} = dimensionless grain diameter (-)
h_0 = initial water depth (before scouring) (m)
v = kinematic viscosity (m²/s)

The following conclusions can be drawn from Equation (5.11). The scour depth is significantly reduced by a substantial decrease in both the jet velocities and the discharge which also follows from Equation (5.9). Moreover, a reduction in the scour depth is obtained when the tailwater depth increases. Furthermore, the impact of the angle on the equilibrium scour depth is marginal for $60° < \theta < 90°$.

Being analogous to that of Schoklitsch (1932), the unknown c_{2V} is related to the particle diameter d_{90}. Hoffmans used over 700 experiments for calibration and validation, including about 50 prototype results (see Hoffmans 2009). The prediction accuracy of the formula is about 93% for a discrepancy ratio r between the predicted and measured value in the range of 0.67–1.5. For coarse gravel and rock (thus for $c_{2V} = 2.9$), the above formula and the formula of Veronese (1937) are comparable:

$$y_{m,e} + h_0 = \alpha_v\, q^{0.54} H^{0.225} \quad \text{with} \quad a_v = 1.9 (\text{not dimensionless!}) \tag{5.12}$$

Furthermore, the scour relation proposed by Schoklitsch (Equation 5.8) yields satisfactory results for both flume and prototype experiments provided $d = 0.025$ m is used for prototype situations. Since the relation of Schoklitsch cannot correct the angle of impingement, its computational results are poor for the experiments of Bormann and Julien (1991).

Based on flume experiments, Breusers and Raudkivi (1991) derived a relation between the equilibrium depth and 3D vertical jets (see also Hoffmans and Verheij 2011):

$$y_{ss} = \frac{\alpha_{RAJ} D U_1}{\sqrt{\Delta g d_{50}}} \qquad (5.13)$$

where
   D = jet diameter (m)
   d_{50} = average grain size (m)
   y_{ss} = equilibrium scour depth (static scouring) (m)
   α_{RAJ} (= 0.3) constant (-)

The ratio between dynamic and static scouring varies between 1.0 and 1.5 (Figure 5.5):

$$1.0 < y_{sd}/y_{ss} < 1.5$$

5.4.3 Discussion

Though variations in the form of equations and coefficients occur, the relations discussed here predict a value in the right order of magnitude. The performance of the scour relations has been studied by Hoffmans (2012). For this purpose, about 700 experiments collected from the literature were applied. The scour relation, as proposed

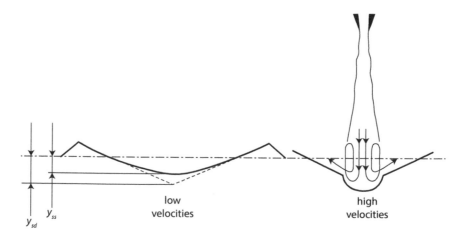

Figure 5.5 Static and dynamic scour.

by Schoklitsch, yields excellent results but only for flume experiments. The relation of Bormann and Julien (1991) can be applied for grade-control structures and also for other types of plunging jets (e.g. classical overflow type). In such cases, the drop height D_p is not taken into account. Overall, the Hoffmans equation (Equation 5.11) being a modified equation of Fahlbusch is doing reasonably well. More than 80% of the computed scour depths fall in the range of 0.5–2.0 times the measured values.

5.5 Two-dimensional culverts

5.5.1 Introduction

The term 'two-dimensional culverts' refers to jets under hydraulic structures that are sufficiently wide. Flow under a gate or barrier at the downstream end of a hydraulic structure or flow out of a rectangular slot has a considerable potential for scour. Jets which discharge entirely under the free surface fall into the class of submerged jets. Many different jet forms can occur, as can be seen in Figure 5.6 (Schoklitsch, 1935, 1962). The form of the scour depends on a number of factors including submergence and the degree of dissipation of the jet energy. Several researchers have investigated the scour caused by a submerged horizontal jet over an erodible bed with and without bed protection, among which Qayoum (1960), Altinbilek and Basmaci (1973) and Rajaratnam (1981).

5.5.2 Calculation methods

Several empirical and semi-empirical relations have been developed for predicting the scour resulting from 2D jets. A comprehensive list of such relations is presented by Whittaker and Schleiss (1984) and Breusers and Raudkivi (1991). Here, some relations are selected which have proven to be of general applicability and to give reasonable results.

Qayoum (1960) studied the scour resulting from flow under gates without bed protection. Several tests were performed in which the discharge, the head and the sediment size varied. Using dimensional analysis, Qayoum obtained the following empirical relation:

$$y_{m,e} + h_t = 2.78 \frac{q^{0.4} H^{0.22} h_t^{0.4}}{g^{0.2} d_{90}^{0.22}} \tag{5.14}$$

where
   d_{90} = particle diameter for which 90% of the mixture is smaller than d_{90} (m)
   h_t = tailwater depth (m)
   H = difference in height between upstream and downstream water levels (m)

Equation (5.14) is based on laboratory experiments but not validated for prototype conditions and thus not generally applicable.

Altinbilek and Basmaci (1973) proposed a method for computing the equilibrium scour depth under vertical gates in a cohesionless bed under the action of horizontal submerged jets. On the basis of 19 model tests with quartz sand and tuff material

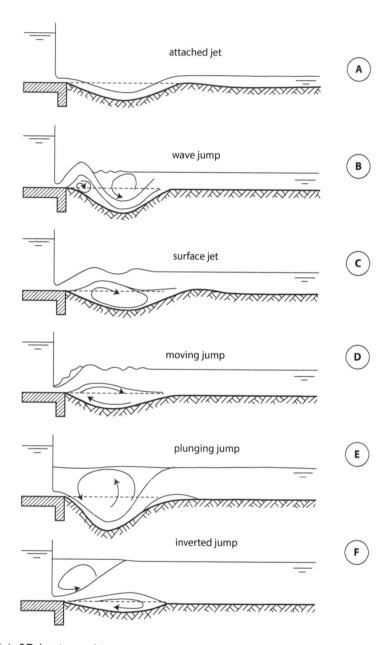

Figure 5.6 **2D horizontal jets.**

(relative density of sand and tuff was 1.6 and 1.3, respectively), the equilibrium scour depth could be closely approximated as follows:

$$y_{m,e} = y_1 \sqrt{\frac{y_1}{d_{50}} \tan \phi} \left(\frac{Fr_1}{\sqrt{\Delta}} \right)^{1.5} \tag{5.15}$$

where
Fr_1 = Froude number (–), $Fr_1 = U_1/(gy_1)^{0.5}$
y_1= thickness of the jet at the vena contracta (m) for
ϕ = angle of repose, $\phi = 40°$

Equation (5.15) is based also on small-scale laboratory tests and not validated for prototype conditions; subsequently, the equation is not generally applicable.

Breusers and Raudkivi (1991) used the main characteristics for fully developed jet flow to describe the dimensions of the scour hole. Based on about 40 flume experiments collected from the literature, they found the following relation submerged horizontal jets:

$$y_{m,e} = 0.008 \, y_1 \left(\frac{U_1}{u_{*,c}} \right)^2 \text{ with } u_{*,c} = \sqrt{\Psi_c \Delta g d_{50}} \tag{5.16}$$

where
$u_{*,c}$ = critical bed shear velocity (m/s)
Ψ_c = critical Shields parameter (–)

These experiments showed that the ratio of length to depth of the scour is approximately constant. The length of the scour hole L_s is roughly 5–7 times the scour depth. Equation (5.16) is a semi-empirical formula. It is based on jet theory and in principle the formula ought to be generally applicable. However, the prediction potential proved to be rather poor (Hoffmans & Verheij, 1997). Hoffmans (2012) describes the scour depth in the equilibrium stage for 2D-H jets (see Figure 5.7). These equations are based on the balance of forces and read:

$$y_{m,e} = c_{2H} \sqrt{q(U_1 - U_2)/g} \text{ with } c_{2H} = \frac{20}{\sqrt{D_{90*}}} \text{ and } D_{90*} = d_{90} \cdot \left(\frac{\Delta g}{v^2} \right)^{1/3} \tag{5.17}$$

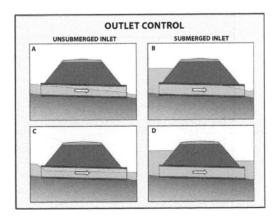

Figure 5.7 **2D** culvert (Source: http://xmswiki.com/wiki/HY-8:Outlet_Control_ Computations).

where

c_{2H} = dimensionless parameter for 2D-H jets (-)
q = discharge in a 2D-H jet (m³/s per meter or m²/s)
U_1 = average flow velocity in the jet (m/s)
U_2 = average flow velocity downstream of the scour hole (m/s)

Equation (5.17) is a semi-empirical formula which approach is based on a balance of forces. Although about 100 flume experiments were used for calibration and validation (see Hoffmans, 2012) the equation is not checked with prototype data. Therefore, it should be applied with care.

For horizontal jets, the hydrostatic force at Section 1 in Figure 5.8 is assumed to be equal to F_2. This assumption is only fair if the flow depth downstream of the hydraulic structure equals approximately the tailwater depth. When the jump is unstable, i.e. when the jump is receding to a point far downstream of the outlet, the assumption $F_1 = F_2$ cannot be applied.

When the bed is protected by an apron and no sediment is supplied from upstream, the equilibrium scour depth can be computed from the Dietz Equation (3.16). Directly downstream of a hydraulic jump, the relative turbulence intensity r_0 ranges from 0.3 to 0.6. The level of turbulence decreases to the normal level of uniform flow at a distance of 20–50 times the flow depth from the end of the hydraulic jump. When the bed protection is longer than the length of the submerged eddy, the relative turbulence intensity can be estimated by (Hoffmans, 1994):

$$r_0 = \sqrt{0.0225 \left(\frac{b_u}{h_t}\right)^{-2} \left(\frac{L - 6(h_t - b_u)}{6.67 h_t} + 1\right)^{-1.08} + 1.45 \frac{g}{C^2}} \quad \text{for} \quad b_u > 0.3 h_t \qquad (5.18)$$

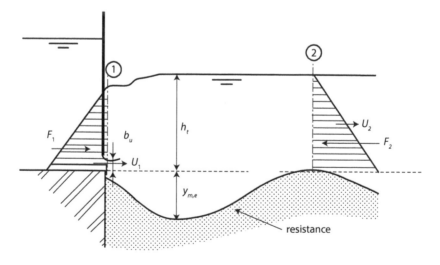

Figure 5.8 Definition sketch of underflow.

where

b_u = thickness of jet at inflow section (m) (Figure 5.8)
C = Chézy coefficient related to bed protection ($m^{1/2}$/s)
L = length of bed protection, $L > 6(h_t - b_u)$ (m)
r_0 = relative turbulence intensity (–)

Equation (5.18) is based on the equations of turbulent kinetic energy transport and dissipation. As a consequence, it is valid in a general sense. However, the equation has not been validated for prototype data and is thus not generally applicable.

Eggenberger and Müller (1944) also studied flow taking place simultaneously over and under a sluice gate. They obtained relations from experiments in laboratory flumes with fine beds, but the accuracy for prototype experiments is questionable.

5.5.3 Discussion

The scour relations discussed here are dimensionally correct. At least 100 experiments on model scale were used to determine the performance of these relations. The relation of Breusers can only be used to estimate scour depth for submerged horizontal jets. For unsubmerged jets, when the tailwater depth is about equal to the jet thickness, the computational results of Equation (5.16) are poor. The methods of Qayoum (1960) and Altinbilek and Basmaci (1973) yield reasonable results for the flume data investigated.

The calibration of c_{2H} in Equation (5.17) was based on approximately 100 flume experiments in which the test section consisted of non-cohesive materials. The hydraulic conditions of the experiments were almost identical and, moreover, no prototype experiments were used. This is a rather narrow basis for assuming that the value of c_{2H} given is the best value. The prediction accuracy of the formula is about 89% for a discrepancy ratio r between the predicted and measured value in the range of 0.5–2.0.

The Dietz Equation (3.16) for the prediction of the equilibrium scour depth based on the continuity approach combined with Equation (5.18) has been successfully verified for about 10 experiments of Shalash (1959). Shalash (1959) studied the influence of apron length L for underflow alone and conducted about 30 experiments in a laboratory flume. In the experiments with a short horizontal bed the minimum value of b_u/h_t was about 0.3. It is therefore necessary to be prudent when extrapolating to prototype situations if b_u/h_t is smaller than 0.3. Finally, regarding the flow pattern in 2D and 3D culverts, the reader is referred to manuals.

5.6 Three-dimensional culverts

5.6.1 Introduction

The prediction of localised scour geometry at circular and square-shaped outlets has long been an element of the culvert design process for determining erosion protection. Three-dimensional culverts are usually designed to carry tributary drainage through roadway embankments. These outlets are defined as 'those devices discharging water where the tailwater depth is usually less than the diameter or the width of the outlet'. However, there are also types of outlet that cause a 3D scour pattern in which the tailwater depth is relatively high.

Figure 5.9 **Examples of culverts.**

Culverts are long structures with a circular or square cross-section connecting flows at both sides of a road or dike (see Figure 5.9). They can also be applied if a canal crosses a small stream or ditch. In principle, the bed of the stream does not continue unlike the bed at a bridge or an aqueduct. Culverts are made of concrete or steel, generally. In the past, culverts were made of bricks, but wood was also used. Specific types of culverts exist, such as a siphon. Easily, a simple sluice can be realised by mounting a valve that allows the water to flow in one direction only.

Scour may occur if the bed directly behind the culverts is unprotected. As scour may undermine the structure itself, the bed in the surroundings of the culvert should be protected sufficiently at both sides. Moreover, it can be necessary to protect the inflow area in case of increased flow velocities due to the constriction (see Figure 5.11). The scour can be computed with the formulas presented in Section 5.7.2; see also the left sketch in Figure 5.10. However, scour may also occur in horizontal direction; see the right sketch in Figure 5.10.

5.6.2 Calculation methods

For most situations, the various relations result in large differences in predicted scour depth. If the tailwater depth is relatively high, a hydraulic jump will be formed (submerged jets). In such cases, the turbulence level is higher and the scouring is most likely more severe than that caused by attached jets (unsubmerged jets) (Figure 5.6).

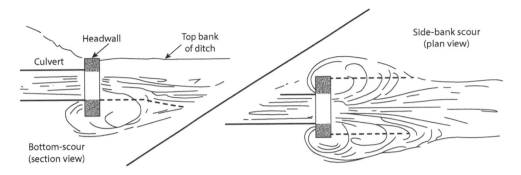

Figure 5.10 Vertical (left) and horizontal (right) scour.

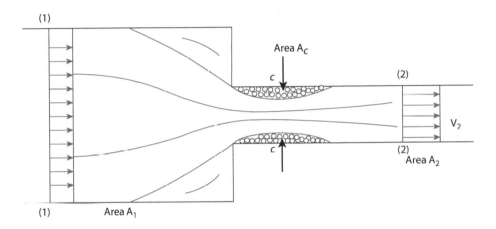

Figure 5.11 Bed protection at the inflow area.

Usually, this phenomenon is not included in the available scour relations. An overview of calculation methods is given by Breusers and Raudkivi (1991). Here, four relations are selected and evaluated. Rajaratnam and Berry (1977) reported results of tests on the erosion of loose beds of sand and polystyrene. The jet velocities were $U_1 = 1.2$–1.8 m/s for sand ($d_{50} = 1.4$ mm) in water and 10–54 m/s in air for polystyrene. With respect to submerged jets, the resulting relation for the equilibrium scour depth is:

$$y_{m,e} = 0.4D\left(\frac{U_1}{\sqrt{\Delta g d_{50}}} - 2\right) \quad \text{for} \quad 2 < \frac{U_1}{\sqrt{\Delta g d_{50}}} < 14 \tag{5.19}$$

The empirical Equation (5.19) is based on laboratory experiments but not validated for prototype conditions and thus not generally applicable.

Ruff et al. (1982) studied the scour process downstream of circular culverts. Over 100 experiments, ranging from 20 to 1000 minutes in duration, were conducted.

The ranges of the systematic tests are listed in Table 5.2. The best correlation was obtained for the relation:

$$y_{m,e} = 2.07D \left(\frac{Q}{\sqrt{gD^5}} \right)^{0.45}$$ (5.20)

where Q is the discharge in metres per second. A few tests by Ruff et al. (1982) of scour over a layer of artificially cohesive sediment indicate that the maximum scour depth may also be expressed by Equation (5.20). This empirical equation can be rewritten to Equation (5.22), meaning that it is generally valid.

Following Breusers and Raudkivi (1991), the length of the scour hole L_s is determined by the condition that the maximum flow velocity in the jet no longer exceeds the critical flow velocity. Combining this assumption and Equation (5.1), the following expression for the equilibrium scour depth can be deduced:

$$y_{m,e} = 0.035D \left(\frac{U_1}{u_{*,c}} \right)^{2/3} \quad \text{for} \quad \frac{U_1}{u_{*,c}} > 100$$ (5.21)

Equation (5.21) is based on the experimental data of Ruff et al. but does not fit well compared to other equations. The accuracy is about 60%. The calibration of Equation (5.21) was based on experiments of Clarke (1962) and Rajaratnam and Berry (1977) with a relatively high tailwater depth. The formula can be rewritten to Equation (5.22) of Hoffmans (2012).

Hoffmans (2012) describes the scour depth in the equilibrium stage for 3D-H jets. This equation is based on the balance of forces and reads:

$$y_{m,e} = c_{3H} \sqrt[3]{Q(U_1 - U_2)/g} \quad \text{with} \quad c_{3H} = \frac{7}{\sqrt[3]{D_{90*}}}$$ (5.22)

where
c_{3H} = dimensionless parameter for 3D-H jets (–)

Other important design parameters with respect to jet scour are the length and the width of the scour hole. Experiments have indicated that the ratio of the scour depth to its length is nearly constant. For both circular and plane jets, the scour length L_s is

Table 5.2 Overview of experimental database (3D horizontal jets)

Investigator	d (mm)	h_t (m)	U_I (m/s)	Q (l/s)	B (m)
Clarke (1962)[a]	0.82–2.4	0.16–0.28	2.11–12.0	0.01–0.49	1.13
Rajaratnam and Berry (1977)[a]	1.4–2.8	0.61	1.28–1.81	0.65–0.92	0.31
Ruff et al. (1982)[b]	0.15–35	0.05–0.20	0.95–5.86	3.11–830	1.22–6.10
Blaisdell and Anderson (1989)[b]	1.8–2.24	0.01	0.32–3.18	0.16–1.62	2.13
Doehring and Abt (1994)[b]	1.86–2.5	0.05	1.22–3.18	10.0–26.0	6.10

[a] Submerged jets.
[b] Unsubmerged jets.

approximately equal to seven times the equilibrium scour depth ($L_s = 7y_{m,e}$). The diameter or width of the scour hole of vertical circular jets can be estimated from $5y_{m,e}$.

5.6.3 Discussion

The calibration of c_{3H} in Equation (5.22) was based on approximately 31 flume experiments with almost identical hydraulic conditions. In the experiments of Clarke (1962) and Rajaratnam and Berry (1977), the hydraulic jump could be considered as submerged. The tailwater depth was relatively large compared to the diameter of the culvert. Subsequently, more than 80 experiments were used to obtain verification of Equation (5.21). In these experiments, the tailwater depth was approximately equal to the jet thickness. In addition, the discharges were, on average, a factor 100 larger and thus almost comparable with prototype conditions. Though the influence of the material properties is marginally taken into account in Equation (5.20), the predictability is reasonable. The prediction accuracy of the formula is about 97% for a discrepancy ratio r between the predicted and measured value in the range of 0.5–2.0 for the 123 tests. The scour relations as proposed by Rajaratnam and Berry (1977) and Breusers and Raudkivi (1991) are based on experiments in which the tailwater depth was relatively high. Therefore, Equations (5.19) and (5.21) can be applied only for submerged 3D horizontal jets. The empirical relation given by Ruff et al. (1982) yields reasonable results on average (accuracy about 80%), but the experiments by Clarke using Equation (5.19) are poor.

In general, the change of momentum per unit of time in a body of flowing water is equal to the resultant of all the external forces that are acting on the body. Despite the simplifications made in applying the momentum principle to a short horizontal reach of a scour hole, reasonable results are obtained for both 2D and 3D jet scour in comparison to those obtained by using other scour relations. However, the scour relations discussed here can only be used for a first approximation of the magnitude of the scour depth in the equilibrium phase, so detailed physical-model studies for a particular design remain important. Furthermore, it must be remembered that the erosion under the action of jets is a dynamic process. The resultant scour depends on the interaction of hydraulic, morphological and geohydrologic conditions. They vary if discharges change in time.

5.7 Ship-induced flow and erosion

5.7.1 Introduction

Ship-induced loads consist of return current velocities, bow and stern waves, and flow velocities from the thrusters (main propeller and bow or stern thrusters) (Figure 5.12). Erosion or scour occurs if the load exceeds a critical value of the bed or bank material. The flow velocities in the return current and the propeller jets can be determined applying the appropriate formulas as mentioned in guidelines and manuals, such as Delft Hydraulics (1988), Schiereck and Verhagen (2012), PIANC report 180 (2015), and the Rock Manual (CIRIA/CUR/CETMEF, 2007). Ship-induced waves may erode the material of the canal or river banks. As waves are not considered in this manual, we will not address ship-induced waves but simply refer to Delft Hydraulics (1988).

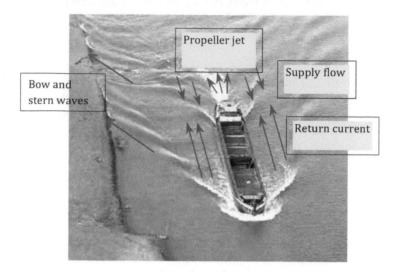

Figure 5.12 **Ship-induced hydraulic loads.**

5.7.2 Scour due to the return current of a sailing vessel

The return current, main propeller jet and bow and stern thruster jets of a sailing vessel may cause erosion of the bed material. This depends on the ratio of the actual and critical flow velocities near the bed. In general, propeller and thruster flow is the dominant factor for erosion at locations where ships accelerate, decelerate or adjust course using stern or bow thrusters, but propeller wash does not play a significant role in straight stretches of a canal where ships normally sail with a speed higher than 1.5 m/s. Nevertheless, as an example that scour due to propeller jets is possible, Figure 5.13 shows a longitudinal erosion pattern resulting from the flow velocities in the main propeller jet for a situation with probably a small keel clearance.

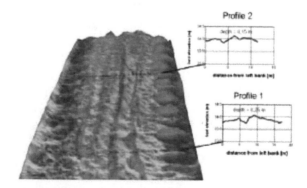

Figure 5.13 **Scour pattern due to a sailing ship (see also Figure 5.14).**

Recently, Deltares and the Dutch Ministry of Public Works carried out a full-scale test in the Juliana Canal in order to predict the erosion of the canal bed by a sailing ship (Dorst et al., 2016). The canal bed consists of gravel. Flow velocities due to the return current were measured below the keel of a passing ship as well as the erosion. Afterwards, the erosion was predicted with an adjusted time-dependent scour formula of Breusers:

$$y_m = (\alpha U - U_c)^{2.0} \sqrt{\frac{t}{K \Delta^{1.7}}} \text{ with } U = U_r + U_0 \text{ and } \alpha = 1.5 + 5 r_0 \tag{5.23}$$

with U_r = ship-induced return current below the ship's keel (m/s), U_0 = natural flow in the canal (m/s) and $K = 330 \, \text{m}^{2.3}/\text{s}^{3.3}$.

Equation (5.23) can be derived with $\gamma = 0.5$ and an exponent 4.0 instead of 4.3 as in the original Breusers formula.

The scour depth as a function of time can be computed with $\alpha = 2.0$ (which implicitly means $r_0 = 0.1$) and:

$$t = Ndt = N \frac{L_s}{V_s} \tag{5.24}$$

where
   L_s = ship length (m)
   V_s = ship speed (m/s)
   N = number of ship passages (-)
   dt = time step (s)

In Figure 5.14, measured and computed results are compared. The computed erosion is roughly twice as large as the measured erosion. This difference could (partly) be caused by the fact that in reality ships do not follow exactly the same course through the canal, contrary to what is assumed in the computation. The tests confirmed that a good estimate of the return current below a ship's keel reads:

$$U_r = (1.5 \text{ to } 2.0) U_{\text{average return flow}} \tag{5.25}$$

With $U_{\text{average return flow}}$ = cross-sectional averaged return current (m/s).

5.7.3 Scour due to propeller and thruster jets

Jets produced by main propellers and thrusters can cause scour at the canal or harbour bed. The scouring effect of the jets is largest when the sailing speed of the vessel is about zero, which occurs during manoeuvring operations when ships use their thrusters for mooring and unmooring. Although the duration of this hydraulic load is very short, the cumulative effect might lead to local scour holes near berths and quay walls. This scouring may result in deformation of the quay wall itself or deformation of structures on the bank (see Figure 5.15).

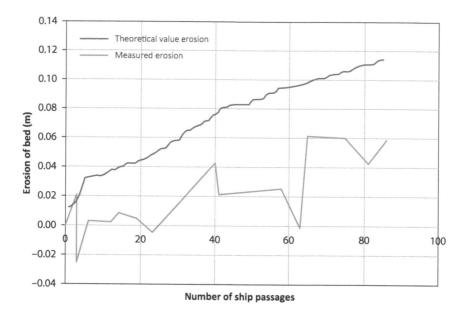

Figure 5.14 Theoretical and measured erosion depths of the canal bottom as a function of the number of ship passages (the theoretical value has been computed using the measured peak velocity of the return current and a critical flow velocity of 0.75 m/s) (Dorst et al., 2016).

Figure 5.15 Local scour in front of a quay wall (left) and deformation of a guard rail (right).

Concerning scour by propeller and thruster jets, four different situations should be distinguished:

1. Horizontal jet parallel to a horizontal bottom;
2. Obliquely downwards oriented thruster jet impinging on a horizontal bottom;
3. Jet impinging on the slope of a bank in a harbour or canal;
4. Jet reflected by a quay wall.

The validity of a specific scour formula is generally restricted to only one of these situations. In all situations, the current velocity just above the non-eroded bottom is the governing factor for scour. In situation 1, also the height of the propeller or thruster jet above the bottom is important. In situation 4, the rate of erosion is larger than in other situations because the volume of the scour hole – given a certain scour depth – is smaller than in the other situations, due to the presence of the quay wall. The situations 2 and 3 are quite similar, but in situation 3, the stability of bottom material is enlarged by the slope angle of the bank of harbour or canal.

Information about scouring caused by propellers can be found in PIANC report 180 (2015) which presents formulas to compute the scour. Furthermore, the time-dependent Breusers formula can be applied and the equilibrium scour depth can be computed; see Section 3.4 (and also Sections 7.2 and 7.3 in Hoffmans (2012) and Hoffmans and Verheij (2011)). See also the method for horizontal 3D jets in Section 5.6.2. However, these formulas should be applied carefully. It is recommended to compare the computed scour with scour at existing locations with comparable conditions. Sections 9.3 and 9.4 present case studies regarding full-scale propeller wash erosion.

Furthermore, Equation (5.20) does not contain the slope angle or parameters describing the scour resistance of the bottom material. As a consequence, the formula is only valid for the investigated case of silty to clayey sand with a slope of 1:4 (Ruff et al., 1982).

5.7.4 Discussion

In general, reliable predictions of the scour depth caused by propeller and thruster jets remain difficult due to the following factors:

- The duration of the hydraulic load on a specific bottom location is very short for a single vessel manoeuvre. The location of the maximum hydraulic load also moves during a manoeuvre.
- The scour depth increases due to the cumulative effect of several manoeuvres at the same location. However, if vessels moor at different locations along a quay wall or jetty, an existing scour hole can partly be filled by bottom material eroded from a new scour hole.
- The hydraulic load on the bottom is not the same for all manoeuvres but varies depending on, for instance, the sizes of the different mooring vessels, the actual draught of vessels, the varying water level in tidal areas, the actual wind condition and the human factor in manoeuvring.
- The engine power actually used is often not well known and can differ between different manoeuvres even of the same vessel (e.g. depending on the wind condition and on the human factor in manoeuvring). The deployment of tugs will reduce the use of particularly the main propeller.
- The scour depth depends on the characteristics of the bottom material. The amount of cohesion is often not well known.
- The frequency of occurrence of the maximum hydraulic load on the bottom can vary strongly between different mooring facilities.

The largest scour depths generally occur at berthing sites of ferries and ro-ro vessels because these vessels always moor at exactly the same location, while the maximum hydraulic load occurs relatively frequently. Recently, a study has been carried out on scour at a quay wall in the Port of Rotterdam. The study showed significant differences between observed scour and scour computed with formulas in PIANC report 180. Sometimes the computed scour is underestimated and in other cases overestimated. A case in Chapter 9 illustrates the problems with predicting the correct scour.

However, the time-dependent development of scour using the Breusers method implicitly assumes that the situation of a propeller at a certain height above the bottom is schematised as an outflow opening at the bottom with an opening height h_{pb} (which is the vertical distance between the propeller axis and the bottom). Obviously, this is not a correct schematisation. Nevertheless application of Breusers for situation 1 can be considered as justified because in situation 1, the maximum flow velocity near the bottom is proportional to h_{pb}, whereas in case of channel flow with a given flow rate, the flow velocity near the bed, $u_{b,max}$, is roughly proportional to the water depth h_0.

In conclusion: PIANC report 180 presents some formulas for scour by propeller and thruster jets in specific situations. These formulas must be applied with great care. This holds also for applying the time-dependent Breusers approach.

5.8 Scour at broken pipelines

Scour due to flow can occur at pipelines and cables on the bottom of a canal or river. Also, the top cover of pipelines and cables can be removed by the water flowing over it. This can result in insufficient stability and subsequent failure if the load increases. Moreover, as a result of the forming of a sill, the scour process can become more serious because the turbulence increases due to the larger eddies. Directly after the sill, a mixing layer will develop with an eddy. This aspect can be treated with the formulas for sills.

Quite different is the development of a broken or leaking pipeline and the subsequent scouring. NEN 3651 (2012) provides safety requirements for pipelines on land and situated in or in the direct surroundings of hydraulic structures.

Hereafter, we discuss the erosion zone (or scour hole) due to a pipeline failure by breaking or leaking. The erosion zone is a soil zone disturbed by the erosive action of escaping liquid, gas or chemicals. Figure 5.16 shows the sink hole created by the breaking of a water pipeline at the VU Medical Centre in Amsterdam.

Figure 5.17 shows the disturbance zone of a broken gas pipeline. The dimensions are: $G_L = 0.5G_B$ for a small hole, $G_L = 2G_B$ for full sliding and outflow at two sides, and $G_L = G_B$ for a large hole or outflow at one side, where G_L and G_B are scour hole length and scour width, respectively. NEN 3651 (2012) makes a distinction between gas and liquid pipelines.

Hoffmans (2012) derived a formula for the equilibrium depth for 3D-H jets on the basis of a balance of forces; see Equation (5.22). This equation can be applied to scour holes due to a broken pipeline.

5.9 Scour control

The relatively high velocities in jet flow, along with the high turbulence level, result in scouring of the river. Scouring can have the following effects.

Figure 5.16 **Sink hole due to a broken pipeline at VU Medical Centre (Foto: NRC / Sam de Voogt).**

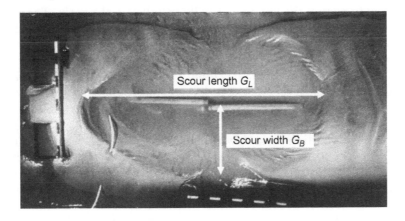

Figure 5.17 **Erosion zone of a gas pipeline.**

- Endangering of the stability of the structure itself by scour holes with initial slopes that are too steep;
- Endangering of the stability of the downstream river bed and banks;
- Formation of a mound of eroded material which may raise the tailwater level.

Therefore, one of the options is to apply a bed protection (see Figure 5.18).

A hydraulic jump is a useful means of dissipating energy in supercritical flow. Its merit is to prevent scouring downstream of dam spillways, weirs, sluices and other

Figure 5.18 **Bed protection downstream of a 3D-H culvert.**

hydraulic structures since it quickly reduces the flow velocities within a relatively short distance. The energy of the hydraulic jump can be dissipated in a stilling basin (Figures 5.19 and 5.20).

The stilling basin is seldom designed to confine the entire length of a free hydraulic jump since such a basin would be too expensive. Consequently, accessories (e.g. baffled aprons) to control the jump are usually installed in the basin. The main purpose of such controls is to shorten the range within which the jump will take place and thus reduce the size and cost of the stilling basin. The hydraulic jump can also be controlled by sills of various designs, such as sharp-crested weirs, broad-crested weirs, and abrupt rises and drops in channel floor. Ski-jump spillways (or bucket-type energy dissipators) are widely used for schemes where, due to the high flow rates and Froude numbers, stilling basins are no longer cost-effective. These devices can be operated either under submerged or under free discharge conditions.

The length of a stilling basin downstream of an overflow structure is based on the dimensions of the structure, the flow phenomena, the geotechnical aspects and safety requirements. Details of the method are presented by Grotentraast et al. (1988). In order to reduce the remaining energy downstream of the stilling basin in most situations, a bed protection is required. Nevertheless, always the geotechnical stability (see Section 3.4) must be checked, taking into account the equilibrium scour depth. The required bed protection length depends on proper functioning of the stilling basin, the water depth, the bed material and the flow velocities. The bed protection should be sufficiently heavy to withstand uplift forces due to water level differences and obviously forces from the flow velocity. Moreover, the bed protection should be sufficiently permeable to relieve leakage water.

In practice, the length of a bed protection varies from about 3 m for small weirs to about 30 m for large weirs. For complex hydraulic structures, other criteria hold and sometimes scale model investigations will be necessary. As a rule of thumb, a first estimate can be a length of at least 15 times the sill height (see Hoffmans 2012). Experimental research has shown that the location of maximum hydraulic load downstream

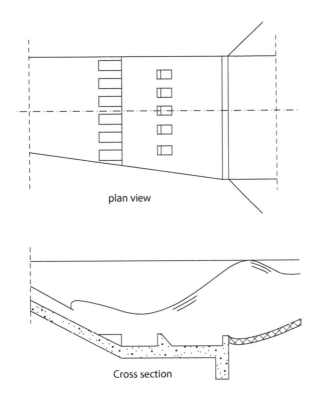

plan view

Cross section

Figure 5.19 **Stilling basin.**

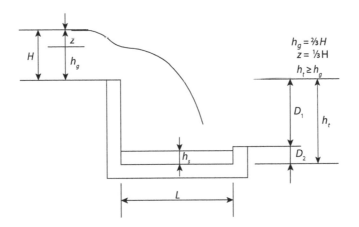

$h_g = \frac{2}{3}H$
$z = \frac{1}{3}H$
$h_t \geq h_g$

Figure 5.20 **Relevant sizes of a stilling basin.**

of a sill with height D lies in the range $10 < L/D < 20$ (with L = length bed protection). Further downstream the hydraulic load decreases gradually till reaching uniform flow conditions. If after some time erosion is observed downstream of the bed protection, it is easy to add protection material. Obviously, monitoring the unprotected bed is essential.

Downstream of culverts, i.e. under underflow conditions, a first estimate of the length of the bed protection should be at least 6 times the water depth minus the outflow height (see Figure 5.8 and Equation (5.18)). In essence, the length of the bed protection must be longer than the length of the hydraulic jump, but if a hydraulic jump occurs, it is recommended to build a stilling basin.

Higher flow velocities and larger eddies may occur at the upstream side of the structure due to the smaller cross-section at the inflow. This can erode bed and banks of the upstream canal. Therefore, it can be important to build a bed protection. The length varies between 1 to 5 m for small structures.

5.10 Examples

5.10.1 Introduction

This section addresses three types of jet scour. Section 5.10.2 shows different computations of the 2D equilibrium scour depth downstream a broad-crested sill with and without a bed protection. Section 5.10.3 discusses how to translate 3D scour downstream a short-crested overflow weir measured in a small-scale experiment to scour in the prototype. Section 5.10.4 shows the computation of 2D scour downstream a gate with under flow. The equilibrium scour depth is determined with various formulas for situations with and without a bed protection.

5.10.2 Two-dimensional scour downstream a broad-crested sill

A barrage is built in a river which has a discharge of $Q = 16\,\mathrm{m^3/s}$. The hydraulic structure can be schematised as a broad-crested sill without bed protection in which the flow above the sill is supercritical. The dimensions of the grade-control structure are: the width is $B = 5\,\mathrm{m}$; the drop height is $D = 2.25\,\mathrm{m}$; and the face angle of the structure is $\lambda = 60°$. The jet angle is about $45°$ and the tailwater depth is $h_t = 3.5\,\mathrm{m}$. The non-cohesive bed material characteristics are $d_{35} = 1850\,\mathrm{\mu m}$, $d_{50} = 2150\,\mathrm{\mu m}$, and $d_{90} = 2500$ μm. The water temperature is 20°C, $v = 10^{-6}\,\mathrm{m^2/s}$. Other data are $\rho_s = 2650\,\mathrm{kg/m^3}$ and $\rho = 1000\,\mathrm{kg/m^3}$.

    a.   What is the maximum scour depth in the equilibrium phase if the downstream scour slope is about 1V:12H?

    b.   What is the maximum scour depth if the bed downstream of the hydraulic structure is protected against current and eddies? The length of the hydraulically rough bed protection is $L = 35\,\mathrm{m}$. Due to diverging walls (1:10), the maximum flow velocity at the end of the bed is about 0.5 m/s and the relative turbulence intensity is estimated to be 0.30.

Solution:

    a.   Maximum scour depth without bed protection
        First, the critical flow depth h_c above the sill for $Fr = 1$ is calculated at:

$$h_c = \left(q^2/g\right)^{1/3} = \left(3.2^2/9.8\right)^{1/3} = 1.0\ \mathrm{m}$$

Then, the mean flow velocity above the sill follows with the continuity equation:

$U_1 = q/h_c = 3.2/1.0 = 3.2$ m/s

The maximum scour depth follows with Equation (5.11):

$$y_{m,e} = c_{2V} \times \sqrt{(q \times U_1 \times \sin q/g)} - h_t = 5\sqrt{(3.2 \times 3.2 \times \sin 45°/9.8)} - 3.5 = 0.8 \text{ m}$$

b.   Maximum scour depth with bed protection
The scour depth will be computed with the Dietz Equation (3.16).
First, relevant parameters will be computed (Figure 5.21):

$$D_* = d_{50} \times (\Delta \times g/v^2)^{1/3} = 2150 \times 10^{-6} \times (1.65 \times 9.8/(1 \times 10^{-6})^2)^{1/3}$$
$$= 54.4 \ \left(\text{with Van Rijn}\right)$$
$$\Psi_c = 0.013 \times D_*^{0.29} = 0.041 \ \left(20 < D_* < 150\right)$$

$$U_c = \sqrt{(\Psi_c \times \Delta \times g \times d_{50})}/\kappa \times \ln\left(12 \times h_t/(3 \times d_{90})\right)$$
$$= \sqrt{(0.041 \times 1.65 \times 9.8 \times 2150 \times 10^{-6})}/0.4 \times \ln\left(12 \times 3.5/(3 \times 2500 \times 10^{-6})\right)$$
$$= 0.82 \text{ m/s}$$

Then substituting all values into the Dietz Equation (5.17) gives:

$$y_{m,e} = h_t \times \left[(1 + 3r_0) \times U_0 - U_c\right]/U_c = 3.5 \times \left[(1 + 3 \times 0.30) \times (0.5) - 0.82\right]/0.82 = 0.55 \text{ m}$$

5.10.3 Three-dimensional scour downstream a short-crested overflow weir

An overflow weir (short-crested sill) is part of a flood storage area in which it was expected that water would spill over the sill very infrequently and, on each occasion, for
a relatively short time. It was decided to use crushed bakelite to model the prototype
bed material, and the model scale was 1/12. Froudian scaling was used to model the
flows. The relevant prototype and model characteristics are listed in Table 5.3.

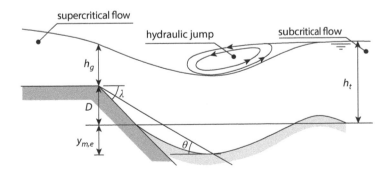

Figure 5.21 Scour hole downstream of a grade-control structure.

Table 5.3 Prototype and model characteristics (Farhoudi & Smith 1982)

	Prototype	Model
h_d (drop height)	2.29 m	0.19 m
q (discharge)	0.37 m^2/s	0.0089 m^2/s
Bed material	Sand	Bakelite
d_{50} (median grain size)	0.12 mm	0.25 mm
Δ (relative density)	1.68	0.40
v (kinematic viscosity)	0.86 × 10^{-6} m^2/s	1.12 × 10^{-6} m^2/s
h_t (tailwater depth)	0.61 m	0.051 m

c. What are the mean velocities at the end of the bed protection in the model scale and in the prototype situation?

d. What are the Froude numbers downstream of the hydraulic jump (model and prototype)?

e. What is the conjugate depth d_1 in the prototype if the Froude number upstream of the hydraulic jump is $Fr_1 = 4.5$ (drowned jump)?

f. What are the critical mean velocities (model and prototype)?

g. What is the time scale factor between model and prototype if $r_{0,p}/r_{0,m} = 1$ ($r_0 = 0.3$)?

h. Experimental results have shown that the maximum scour depth in the model is about 0.1 m after 2 hours ($y_m = d_0$). What is the maximum scour depth in the prototype situation after 36 hours?

Solution:

a. Mean velocities
    Model:

$U_0 = q/h_t = 0.0089/0.051 = 0.175$ m/s

Prototype:

$U_0 = q/h_t = 0.37/0.61 = 0.607$ m/s

b. Froude numbers
    Model:

$Fr = U_0/(g \times h_0)^{1/2} = 0.175/(9.8 \times 0.051)^{1/2} = 0.25$

Prototype:

$Fr = U_0/(g \times h_0)^{1/2} = 0.61/(9.8 \times 0.61)^{1/2} = 0.25$

c. Conjugate depth
    Prototype:

$$d_1 = \left[q/(g^{0.5} \times Fr_1) \right]^{2/3} = \left[0.37/(9.8^{0.5} \times 4.5) \right]^{2/3} = 0.09 \text{ m}$$

d. Critical mean velocities (van Rijn)
   Model:

$$D_* = d_{50} \times (\Delta \times g/v^2)^{1/3} = 250 \times 10^{-6} \times \left(0.40 \times 9.8/(1.12 \times 10^{-6})^2\right)^{1/3} = 3.66$$
$$\Psi_c = 0.24 \times D_*^{-1} = 0.0656 (D_* < 4)$$
$$U_c = \sqrt{(\Psi_c \times D \times g \times d_{50})}/\kappa \times \ln(12 \times h_t/(3 \times d_{90}))$$
$$= \sqrt{(0.0656 \times 0.4 \times 9.8 \times 250 \times 10^{-6})}/0.4 \times \ln\left(12 \times 0.051/(3 \times 250 \times 10^{-6})\right)$$
$$= 0.13 \text{ m/s}$$

Prototype:

$$D_* = d_{50} \times (\Delta \times g/v^2)^{1/3} = 120 \times 10^{-6} \times \left(1.68 \times 9.8/(0.86 \times 10^{-6})^2\right)^{1/3} = 3.38$$
$$\Psi_c = 0.24 \times D_*^{-1} = 0.0711 (D_* < 4)$$
$$U_c = \sqrt{(\Psi_c \times D \times g \times d_{50})}/\kappa \times \ln(12 \times h_t/(3 \times d_{90}))$$
$$= \sqrt{(0.0711 \times 1.68 \times 9.8 \times 120 \times 10^{-6})}/0.4 \times \ln\left(12 \times 0.61/(3 \times 120 \times 10^{-6})\right)$$
$$= 0.29 \text{ m/s}$$

e. Scale factor between model and prototype
   Equation (5.7):

$$n_t = (h_{t,p}/h_{t,m})^2 \times (\Delta_p/\Delta_m)^{1.4} \times \left(Fr_p/Fr_m\right)^{0.9} \times \left[(a_F \times U_0 - U_c)_p/(a_F \times U_0 - U_c)_m\right]^{-3}$$
$$= (0.61/0.051)^2 \times (1.68/0.4)^{1.4} \times \left[(1.15 - 0.29)/(0.33 - 0.13)\right]^{-3}$$
$$= 13.4$$

This means, for example, that 1 min of model time represents approximately 13 min in prototype time.

f. Maximum scour depth after 36 hours
   Model:

$$t_{1,m} = 2 \text{ hours}$$
$$d_{0,m} = 1/2 \times h_{d,m} = 1/2 \times 0.19 = 0.095 \text{ m}$$

Prototype:

$$t_{1,p} = n_t \times t_{1,m} = 13.4 \times 2.0 = 26.8 \text{ hours}$$
$$d_{0,p} = 1/2 \times h_{d,p} = 1/2 \times 2.29 = 1.145 \text{ m}$$

Equation (5.13):

$$y_{m,p} = d_{0,p} \times \left(t/t_{1,p}\right)^{0.2} = 1.145 \times (36/26.8)^{0.2} = 1.2 \text{ m}$$

After 36 hours, the maximum scour depth in the prototype is about 1.2 m.

5.10.4 Two-dimensional scour downstream an under flow gate

In a river, a gate is built which operates as a sluice gate (Figure 5.22). For $q = 2 \text{ m}^2/\text{s}$, the upstream water level is 10 m. The tail characteristics are $h_t = 5 \text{ m}$ and the jet thickness is 0.34 m. The flow is assumed to be 2D. It is also given that $d_{50} = 5 \text{ mm}$, $d_{90} = 7 \text{ mm}$, $\rho_s = 2650 \text{ kg/m}^3$, and $\rho = 1000 \text{ kg/m}^3$.

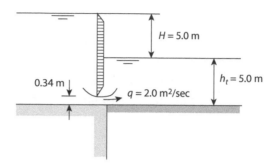

Figure 5.22 **Flow condition at a sluice gate.**

a.   What is the maximum scour depth downstream of the sluice gate according to the methods of Qayoum, Altinbilek, and Breusers?

b.   What is the maximum scour depth if the bed is protected ($L = 75$ m)?

Solution:

a.   Maximum scour depth without bed protection
Qayoum:

$$y_{m,e} = 2.78 \times q^{0.4} \times H^{0.22} \times h_t^{0.4}/(g^{0.2} \times d_{90}^{0.22}) - h_t$$
$$= 2.78 \times 2^{0.4} \times 5^{0.22} \times 5^{0.4}/(9.8^{0.2} \times 0.007^{0.22}) - 5$$
$$= 14 \text{ m}$$

Altinbilek:

$$y_{m,e} = y_1 \times \sqrt{(y_1 \times \tan \phi / d_{50})} \times \left(Fr/\sqrt{\Delta}\right)^{1.5}$$
$$= 0.34 \times (0.34 \times \tan 40°/0.005) \times (3.23/\sqrt{1.65})^{1.5}$$
$$= 10 \text{ m}$$

Breusers:

$$y_{m,e} = 0.008 \times y_1 \times (U_1/u_{*,c})^2 = 0.008 \times 0.34 \times (5.9/0.067)^2 = 21 \text{ m}$$

Since the range of predicted scour depths is relatively large, the aforementioned equations can only be used for first estimates. The order of magnitude of the equilibrium scour depth is 10 m.

b. *Maximum scour depth with bed protection*
Jet thickness at the inflow section:

$$b_u = y_1/m = 0.34/0.59 = 0.58 \text{ m}$$
$$D_* = d_{50} \times (\Delta \times g/v^2)^{1/3} = 5000 \times 10^{-6} \times \left(1.65 \times 9.8/(1 \times 10^{-6})^2\right)^{1/3} = 126 \text{ (van Rijn)}$$
$$\Psi_c = 0.013 \times D_*^{0.29} = 0.053(20 < D_* < 150)$$
$$U_c = \sqrt{(\Psi_c \times \Delta \times g \times d_{50})}/\kappa \times \ln(12 \times h_t/(3 \times d_{90}))$$
$$= \sqrt{(0.053 \times 1.65 \times 9.8 \times 5000 \times 10^{-6})}/0.4 \times \ln\left[12 \times 5/(3 \times 7000 \times 10^{-6})\right]$$
$$= 1.3 \text{ m/s}$$

Equation (5.18):

$$r_0 = \sqrt{\left[0.0225 \times \left(b_u/h_t\right)^{-2} \times \left[\left(L - 6 \times (h_t - b_u)\right)/(6.67 \times h_0) + I\right]^{-1.08}\right]}$$

$$= \sqrt{\left[0.0225 \times (0.58/5)^{-2} \times \left[(75 - 6 \times 4.42)/(6.67 \times 5)\right] + 1\right]^{-1.08}} = 0.80$$

Note that the ratio between the jet thickness at the inflow section and the tailwater depth ($b_u/h_t = 0.12$) is smaller than 0.3. If $b_u/h_t < 0.3$ predictions by Equation (5.18) might give conservative values for r_0.

Assuming that $r_0 = 0.8$ and applying the Dietz method the equilibrium scour depth is:

$$y_{m,e} = h_t \times \left[(1 + 3r_0) \times U_0 - U_c\right]/U_c = 5 \times \left[(1 + 3 \times 0.8) \times 0.4 - 1.3\right]/1.3 = 0.23 \text{ m.}$$

Chapter 6

Sills

6.1 Introduction

In an estuary or a river, a sill may be the initial foundation or the lower part of a structure that has to be constructed on a bed of alluvial materials. A sill is a horizontal, structural part of a structure near the bed level on a foundation or pilings or lying on the ground in earth-fast construction. Sometimes a sill is used to reduce the mixing of different types of water in an estuary. In an estuary, a sill has to be designed for flow in two directions: flood flow and ebb flow. In rivers, for example, a sill may be used as part of a scheme to maintain a minimum water level. Figure 6.1 shows example sketches with characteristics of sills in a river.

Relatively small weirs can be found in agricultural areas (see Figure 6.2). The weir itself is not the subject of this manual but the sill on which it is built. Different types can be distinguished (Grotentraast et al., 1988):

  a.   Fixed weirs: horizontal crest and other crest shapes;
  b.   Adjustable weirs: beam weir, weir with a lift slide, and weir with a valve;
  c.   Specific weirs (fixed and adjustable): segment weir, cascade weir, siphon weir, weir with a counter weight, and inlet or outlet weir.

The flow pattern downstream of sills is discussed in Section 6.2. Section 6.3 summarises relations for the maximum scour depth at sills with horizontal bed protection (Figure 6.1). The upstream scour slope and the gradual undermining at the end of the bed protection due to high turbulence level are treated in Section 6.4. Section 6.5 deals with additional measures to reduce the risk of failure, and Section 6.6 describes some field experiments with loosely packed material. Finally, an example is discussed in Section 6.7.

6.2 Flow characteristics

The bed in the direct neighbourhood of hydraulic structures is generally protected against currents and eddies. The length of the bed protection depends on the permissible scour (permissible maximum scour depth and upstream scour slope) and the geotechnical conditions of the soil involved (e.g. densely or loosely packed sand). For longer bed protections, the scour process is less intense due to the decay of turbulent energy and the adaption of the velocity profile downstream of the hydraulic structure. However, the measures required for protection against scouring are costly,

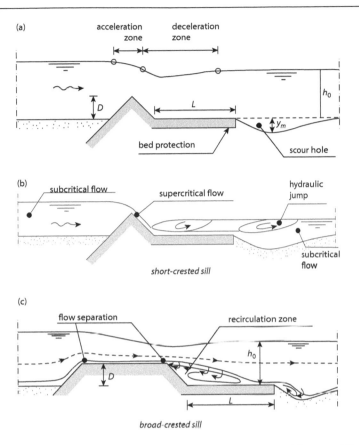

Figure 6.1 **Example sketches of a short-crested sill (top: submerged flow; middle: free flow) and a broad-crested sill designed for one flow direction (left to right).**

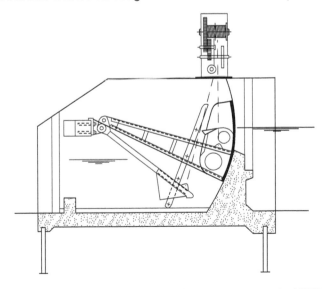

Figure 6.2 **Example of a small segment weir (Grotentraast et al., 1988).**

especially when the bed protection is constructed under water. As an example of the two-dimensional (2D) (width-averaged) flow pattern downstream of a hydraulic structure, Figure 6.3 shows the width-averaged subcritical flow at a sill. Downstream of a hydraulic structure and inside a scour hole, the separated shear layer appears to be similar to a simple plane mixing layer. Initially, the lower bound of the mixing layer is slightly curved due to the influence of the bed.

The curvature increases in the downstream direction; especially, near the point of reattachment (at a distance of about six times the sill height from the sill). Directly behind the sill a reverse flow develops with a direction opposite to the main flow. In both the mixing layer and the recirculation zone, the turbulent energy is higher than in uniform flow conditions.

Vortices with a vertical axis will occur when the flow pattern is influenced by vertical ends or wing walls. The rotating ascending current in the vortex then picks up the bed material and throws it out sideways. The vortex street may become so intense that it endangers the stability of the structure unless effective protective measures are taken. The flow in a scour hole is more or less 2D when a vortex street influences the scour process only marginally, i.e. when the velocity gradients in the transverse direction are relatively small (see Figure 6.4).

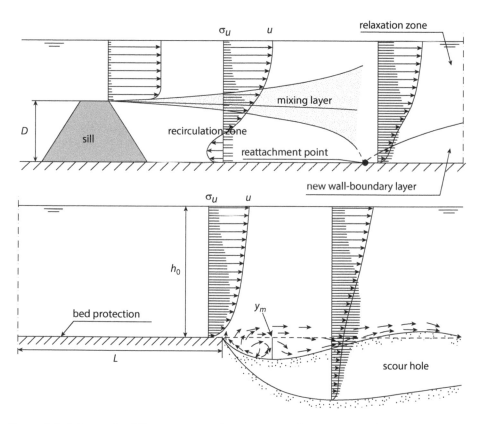

Figure 6.3 Schematised flow patterns downstream of a sill with a full bed protection (top) and a transition from a bed protection to an alluvial bed.

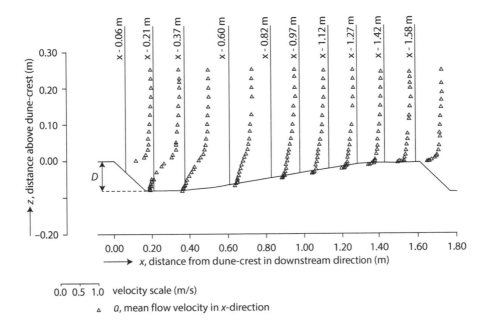

Figure 6.4 $\bar{u}$ as a function of x and z above a dune (T6) (van Mierlo & de Ruiter, 1988).

Downstream of sills and weirs, areas with deceleration and acceleration occur. Between these areas flow characteristics differ, such as main flow area, mixing layer depth, eddy hydraulic jump, core zone of a jet, etc. In order to prevent large scour due to the high flow velocities in jets and the high turbulence intensities downstream of weirs and sills, usually a bed protection is constructed. The bed protection can be combined with a stilling basin in which the energy of the fast flowing water is dissipated.

6.3 Scour depth modelling in the Netherlands

6.3.1 Introduction

For a designer, the most important scour parameter is the maximum scour depth in the equilibrium phase. However, in deltaic areas, generally characterised by large flow depths, the time factor is also important, especially during the closure of estuary branches. The maximum scour depth depends on the bed shear stress and the turbulence condition near the bed, and on sediment characteristics (density of the bed material, sediment size distribution, porosity and cohesiveness). The sediment and flow characteristics determine the type of the scour, i.e. (1) clear-water scour because the bed shear stress upstream of a scour hole is smaller than the critical bed shear stress of the bed material, or (2) live-bed scour when sediment is transported from upstream to the scour hole (Section 3.4.4).

Several hundreds of tests in which no sediment was transported from upstream were carried out at Delft Hydraulics (Delft Hydraulics, 1972, 1979; Buchko, 1986; Buchko et al., 1987). In general, the experimental relations derived from the results at clear-water scour are considered to be reasonably valid (Figure 6.5). However, even with an

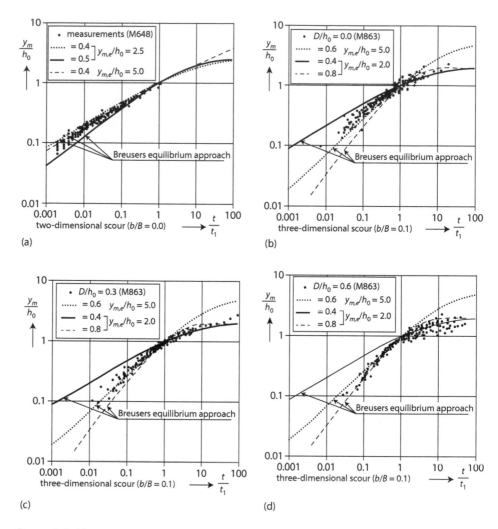

Figure 6.5 Maximum scour depth as a function of time for experiments with clear-water scour.

upstream flow velocity larger than the critical flow velocity, situations are possible that there is no sediment supply from upstream because the flow velocity is too low to create moving bed conditions. This situation occurs for example in the Dutch tidal rivers. During the flood, hardly any sediment is transported from the sea, and during ebb, hardly any sediment is transported to the sea. Subsequently, the scour capacity is larger than under Live Bed Scour (LBS) conditions. In practice, this means that the measured scour depth in a flume is larger than in reality because the scour capacity is larger.

Local scour results directly from the impact of the structure on the flow. Physical model testing and prototype experience have permitted the development of methods and formulas for predicting and preventing scour at different types of structures. Additional experiments in a laboratory may be particularly appropriate for unusual structures not covered by existing formulae or if the risks of scouring are too big for the hydraulic structure. Numerical models enable to predict the flow field and the turbulence at the hydraulic structure (see Section 3.2). This information can be used for the determination of the scour depth.

6.3.2 Scour depth formula

The scour process as a function of time can be given with reasonable accuracy, provided the equilibrium scour depth $(y_{m,e})$ is known:

$$\frac{y_m}{y_{m,e}} = 1 - e^{-\frac{h_0}{y_{m,e}}\left(\frac{t}{t_1}\right)^{\gamma}} \tag{6.1}$$

where
t = time (s)
t_1 = characteristic time (s) at which $y_m = h_0$
h_0 = initial water depth (m)
y_m = maximum scour depth at t (m)
γ = coefficient ( )

In the development phase, when t is smaller than t_1, Equation (6.1) reduces to:

$$\frac{y_m}{h_0} = \left(\frac{t}{t_1}\right)^{\gamma} \tag{6.2}$$

Breusers (1966) validated Equation (6.2) extensively in a systematic research project. In the laboratory experiments, it was shown that no scale effects occurred and thus the equation can also be applied in prototype conditions. Breusers (1966) reported that for 2D scour the averaged value of the coefficient γ amounted to 0.38. According to Mosonyi and Schoppmann (1968) and Dietz (1969), the coefficient γ lies within the range of 0.25–0.40, which can be considered as a confirmation of Breusers' results (Table 6.1).

In 3D situations, for example scour at one end of a sill, a vortex street (vortices with a vertical axis) occurs (Figure 6.6). In this zone, the flow is more turbulent and large vortices intermittently erode and transport bed material. For 3D flow, γ is strongly dependent on the degree of turbulence generated by the vortex street. Figure 6.5

Table 6.1 Coefficient γ

Investigator	γ	Flow condition
Breusers (1966)	0.38	Two-dimensional
Mosonyi and Schoppmann (1968)	0.27–0.35	Two-dimensional
Dietz (1969)	0.34–0.40	Two-dimensional
van der Meulen and Vinjé (1975)	0.4–0.8	Three-dimensional

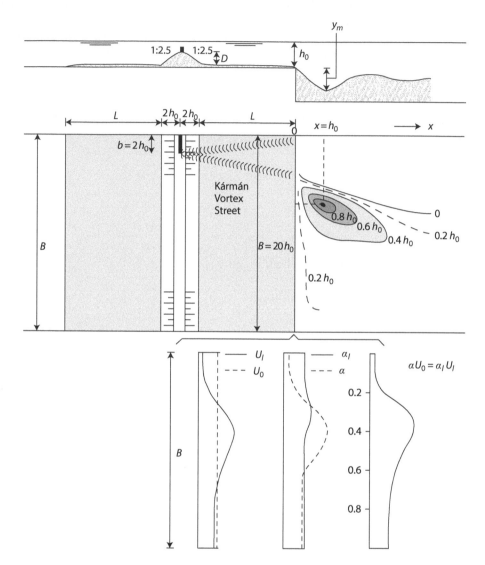

Figure 6.6 **Definition sketch of U_l.**

shows the results of about 110 tests from the systematic research (van der Meulen & Vinjé, 1975), as a function of y_m/h_0 versus t/t_1 for the deepest points of the scour hole in the cross sections ($D/h_0 = 0.0$, 0.3 and 0.6). The scatter in the results is generally within reasonable limits, so that the scour patterns were similar in shape as the scour progressed. However, a different coefficient γ was required for the prediction of the maximum scour depth as a function of time.

For small scour depths, the exponential function of the form of Equation (6.2) gave a good fit if γ values larger than 0.38 were used. In the initial and development phases, a conservative value for the maximum scour depth is obtained if γ is 0.4 for two as well as three-dimensional (3D) situations. In the equilibrium phase, the maximum scour depth approaches a limit. A relatively long period of time is needed to attain equilibrium conditions, especially for experiments with small Froude numbers and no upstream sediment supply (clear-water scour). Therefore, the equilibrium situation was not always achieved with the scale models (Delft Hydraulics, 1972).

Hoffmans (2012) derived new formulas based on an analysis of equilibrium scour depths and the erosion law of Partheniades. The equilibrium scour depth for clear-water scour is (see also Section 3.4 and Table 3.1):

$$\frac{y_{m,e}}{h_0} = 1.0 \chi_e \left(\frac{U_0}{U_c}\right)^2 - 1.0 \quad \text{for} \quad U_0 < U_c \tag{6.3}$$

and for live-bed scour

$$\frac{y_{m,e}}{h_0} = 1.0 \chi_e - 1.0 \quad \text{for} \quad U_0 \geq U_c \tag{6.4}$$

where

$$\chi_e = \frac{1 + 6.3 r_0^2}{1 - 6.3 r_{0,m}^2} \tag{6.5}$$

where
  $y_{m,e}$ = equilibrium scour depth (m)
  h_0 = initial water depth (m)
  r_0 = depth-averaged relative turbulence intensity at water depth change (–)
  $r_{0,m}$ = depth-averaged relative turbulence intensity when scour depth is maximal (–)
  U_0 = depth-averaged flow velocity above the bed (m/s)
  U_c = critical depth-averaged flow velocity above the bed (m/s)
  χ_e = turbulence parameter (–)

In the deepest part of the scour hole, $r_{0,m}$ can be approximated by:

$$r_{0,m} = \sqrt{\tfrac{1}{2} C_k} \left(\frac{y_m}{h_0} + 1\right) \quad \text{for} \quad 0.1 < \frac{y_m}{h_0} < 2 \tag{6.6}$$

where
  C_k = constant dependent on the steepness of the upstream slope, $C_k = 0.03 - 0.045$

As a first estimate for the equilibrium scour depth, a default value of 0.25 can be used for $r_{0,m}$.

Although Dietz (1969) also derived a comparable formula for the equilibrium scour depth (see Equation 3.16) the effects of turbulence in the scour hole itself have not been taken account.

6.3.3 Characteristic time

On the basis of dimensional considerations, the time needed to erode a volume V per unit width could be given by $t = V/s$, where the area of the longitudinal scour section is related to the maximum scour depth ($V = c_a \, y_m^2$, where c_a is a geometrical factor and s is the sediment transport). Several researchers have investigated the dependence of the characteristic time on hydraulic conditions and material characteristics. From model tests, it appeared that for steady flow, the influence of the various parameters could be described by:

$$t_1 = \frac{Kh_0^2\Delta^{1.7}}{(\alpha U_0 - U_c)^{4.3}} \tag{6.7}$$

where
   $K = 330\,\text{m}^{2.3}/\text{s}^{3.3}$, t_1 in hours (or $1.19 \times 10^6\,\text{m}^{2.3}/\text{s}^{3.3}$, t_1 in s)
   α = coefficient depending on the flow velocity and turbulence intensity (–)
   Δ = relative mass density of bed material (–)

Equation (6.7) has been validated extensively during the systematic research programme carried out by Breusers but has been validated also for two prototype situations (Brouwersdam). It is recommended to adjust the description of the sediment transport in order to fulfil the Partheniades erosion law. This results in a different numerator (and of course in adjusted exponents and coefficients too).
   The characteristic time was originally expressed in hours, and the coefficient K is not dimensionless. Using the invariables g (acceleration of gravity) and v (kinematic viscosity), Equation (6.7) can be rewritten as:

$$t_1 = \frac{h_0}{U_0} \frac{K_1\Delta^a}{\alpha_u^b \text{Fr}^c \text{Re}^d} \tag{6.8}$$

where
   $Fr = U_0/(gh_0)^{1/2}$, Froude number (–)
   $K_1 = K/(g^{1.43}v^{0.43})$ (K in $\text{m}^{2.3}/\text{s}^{3.3}$) coefficient (–)
   $Re = U_0h_0/v$, Reynolds number (–)
   v = kinematic viscosity
   α_u = coefficient (–), to be determined as $\alpha_u = \alpha - U_c/U_0$

Dietz (1969) performed extensive research on 2D scour downstream of horizontal beds and low sills. Several non-cohesive materials (sand, lignite and polystyrene) were used in the experiments on model scale in which the initial flow depth varied from 0.125 to 0.25 m. The research of Dietz (1969) confirmed the considerations of Breusers. Although the relation is identical, Dietz proposed different values for the empirical coefficient K_1 and its exponents (Table 6.2).
   Hoffmans (1992) showed that the scour relation based on the sediment transport presented by Zanke (1978) can be written into the form of Equation (6.8). However, the coefficient K is not constant but a weak function of the sediment diameter. Van der Meulen and Vinjé (1975) studied the 3D scour process downstream of a partial channel constriction and reported that K was equal to 250 (based on more than 100 tests with hydraulically smooth, medium and rough bed protections). They concluded that the shape of the scour hole is independent of the bed material and flow velocity, and that Equation (6.9) is equally applicable to 3D situations, provided α is assigned correctly. For 3D scour, the characteristic time is not constant at each cross section, owing to the

Table 6.2 Empirical coefficients and exponents in Equation (6.8)

Investigator	$K_1/10^6$	a	b	c	d
Breusers (1966)	0.94	1.62	4.0	2.7	0.3
Dietz (1969)	9.96	1.5	4.0	2.5	0.5
van der Meulen and Vinjé (1975)	12.9	1.7	4.3	2.87	0.43
Zanke (1978)		1.33	4.0	2.67	0.33
de Graauw and Pilarczyk (1981)	17.1	1.7	4.3	2.87	0.43

3D character of the scour hole. Therefore, it is necessary to determine the development of scour in many longitudinal sections separately.

De Graauw and Pilarczyk (1981) found that K was equal to 330 for tests with rough bed protection. In addition, they specified the factor α for both 2D and 3D flow at the location where the scour depth is at its maximum. The influence of the roughness of the bed protection on the characteristic time t_1 can be taken into account by fitting either the coefficient K or the flow and turbulence factor α. De Graauw (1981) analysed the predictability of t_1 by applying 150–200 tests in a scale model and found that the relative standard deviation $\sigma(t_1)$ was about 30%. The accuracy of the computed y_m (maximum scour depth) with Equation (6.2) is characterised by a relative standard deviation of about 10%.

6.3.4 Relative turbulence intensity

To analyse the decay of turbulence in the relaxation zone, an analogy with the decay of turbulent energy and the dissipation of grid turbulence can be used (Launder & Spalding, 1972). This analogy is shown for the situation where downstream of a sill a scour hole has developed. If the geometry around the sill consists of a horizontal bed and the flow is subcritical above the sill, a relation for the relative turbulence intensity r_0 can be deduced for that situation. In the case without a bed protection downstream of the sill, a scour hole will develop directly downstream. The maximum value of r_0 at the upstream edge of the scour hole is (Hoffmans & Booij, 1993a):

$$r_0 = \sqrt{0.0225\left(1-\frac{D}{h_0}\right)^{-2}+1.45\frac{g}{C^2}} \qquad (6.9)$$

In the same situation, with the sill and in the case with a bed protection with a length along the channel L longer than six times the sill height, the relation is at the upstream edge of the scour hole (end of bed protection):

$$r_0 = \sqrt{0.0225\left(1-\frac{D}{h_0}\right)^{-2}\left(\frac{L-6D}{6.67h_0}+1\right)^{-1.08}+1.45\frac{g}{C^2}} \text{ for } L>6D \qquad (6.10)$$

where
   C = Chézy coefficient related to bed protection ($m^{1/2}/s$)
   D = height of the sill (m)
   g = acceleration of gravity, $g = 9.8\,m/s^2$
   L = length of bed protection ($L > 6D$) (m)

Downstream of the reattachment point the critical shear stress increases up to the equilibrium value and hence r_0 reaches its maximum value $r_0 = 1.2g^{0.5}/C$. The term $1.45g/C^2$ is the maximum value and addresses the uniform flow conditions.

Equation (6.10) is based on equations of turbulent kinetic energy transport and dissipation. The equation has been calibrated and validated with small-scale flume experiments. As the formula has a theoretical background, it can also be applied in prototype conditions. Equation (6.10) is almost identical to Equation (5.18). The height of the submerged eddy is replaced by the height of the sill.

Recently, Arcadis developed an approach to estimate the energy loss downstream of a sill (Hoeve et al., 2015, 2018; Koote, 2017). The approach models the turbulent kinetic energy production and dissipation. The method is interesting because it is easy to apply and potentially allows a reduction of the stone grading of the bed protection downstream of the sill as function of the distance to the sill. However, the method requires further improvement and validation; therefore, it is recommended to apply this method carefully.

The length of a safe bed protection will always have to be extended beyond the point of reattachment. Although some characteristics of 3D flow and additional phenomena, such as vortices with a vertical axis, have not been taken into account in Equations (6.9) and (6.10), promising results for 3D flow fields have been obtained (Figure 6.7).

6.3.5 Scour coefficient

The magnitude of the flow and turbulence coefficient α for local scour depends largely upon the upstream geometry and can be interpreted as a measure of the erosion capacity. For a storm surge barrier α is determined for the greater part by the height of the sill (relative to the initial flow depth) and for a dam with a spillway in the presence or absence of energy dissipators. The length and roughness of the bed protection downstream of the structure also play an important role in the determination of α. Several empirical expressions for α have been deduced from the tests in the systematic series for both 2D and 3D scour (de Graauw, 1981; de Graauw & Pilarczyk, 1981).

According to Breusers (1966, 1967), turbulence in a scour hole plays an important role in the scour process and fully depends on the flow velocity and turbulence level at the transition between the fixed and the erodible bed ($\alpha_l = (1 + 3r_0)c_v$, where c_v is a correction factor for the velocity profile). In 3D situations, the factor α depends on the turbulence level which in turn depends on the 3D flow pattern (Figure 6.6). Jorissen and Vrijling (1989) reported that for hydraulically rough flow conditions, the turbulence effect α_l, can be expressed by:

$$\alpha_l = 1.5 + 5r_0 \tag{6.11}$$

Equation (6.11) has been derived and validated in the systematic scour research program.

A more general relation for the local turbulence coefficient is (Hoffmans & Booij, 1993a):

$$\alpha_l = 1.5 + 4.4r_0 f_C \tag{6.12}$$

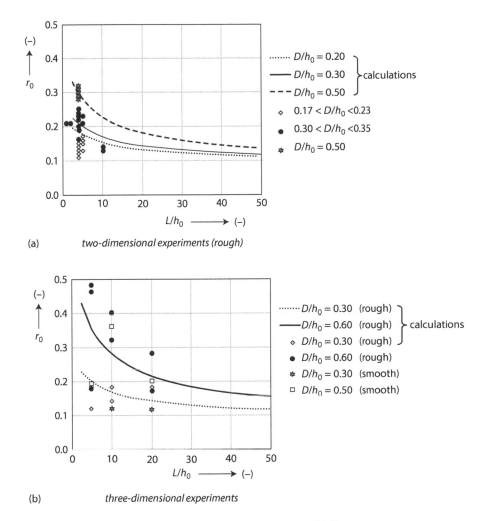

Figure 6.7 Relative turbulence intensity as a function of L/h_0.

where
  f_C = roughness function related to bed protection, $f_C = C/C_0$
  C = Chézy coefficient related to the bed protection upstream from the scour hole
  $C_0 = 40\,\mathrm{m}^{1/2}/\mathrm{s}$, if $C < C_0$, then $f_C = 1$

Equation (6.12) has also been validated in the systematic scour research programme but contains an additional parameter to account for hydraulically smooth beds. For $C = 45\,\mathrm{m}^{1/2}/\mathrm{s}$, Equation (6.12) reduces to Equation (6.11). Equation (6.12) was based on about 250 measurements in a 2D and 3D flow pattern for $r_0 > 0.05$. For 2D flow, the local depth-averaged flow velocity equals approximately the mean flow velocity ($U_l \approx U_0$ thus $\alpha_l \approx \alpha$), whereas for 3D flow U_l and α_l depend strongly on the geometry upstream of the scour hole. Consequently, the development of the scour process in a 3D scour hole can be determined at any cross section, provided α_l and U_l are known variables. Note that α is linked to U_0.

Combining Equations (6.10) and (6.12), the local turbulence coefficient can be given as a function of the geometry upstream of the scour hole. Figure 6.8 shows α_l (α for 2D flow) as a function of L/h_0 for several heights of the sill. Hence, α increases with sill height, which demonstrates that scour increases owing to flow deceleration downstream of the sill. The value of α may be reduced by either lengthening the bed protection or by making the bed rougher. The flow pattern downstream of a sill combined with a horizontal constriction is 3D, at least to some degree, due to the vortex street. In such cases, α is significantly greater than the value of α_l for 2D flow.

When no information is available with respect to the 3D flow pattern, we recommend using design curves presented by de Graauw and Pilarczyk (1981) (Figure 6.8). These curves are valid for $D/h_0 < 0.8$, $L/h_0 > 5$ and $b/B = 0.1$, where b is the width of the abutment and B is the width of the flow. A smooth bed protection results in a faster scour process because the flow above a smooth bed has greater momentum than the flow above a rough bed. For a hydraulically smooth bed, α has to be increased by a term that lies in the range of 0.3–0.5 ($\alpha_{\text{smooth}} = \alpha_{\text{rough}} + 0.4$).

In this section and the foregoing sections, it is assumed that the main flow is subcritical (Froude number is smaller than 1) during all construction stages of the closure. In an estuary, this will be often the case, but supercritical flow is possible when closing a river branch.

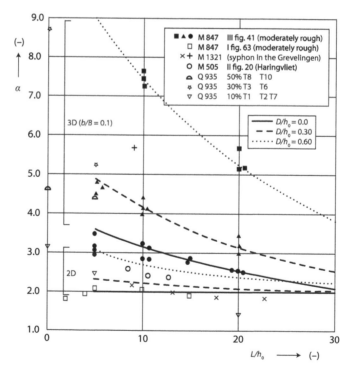

Figure 6.8 α as a function of L/h_0 for 2D scour (below; for $D/h_0 = 0.0$ α equals 2.0) and for 3D scour (top).

6.3.6 Non-steady flow

The following aspects should be studied in order to estimate the scour depth in tidal areas:

- The variation of the discharge in time, which is influenced by tides (daily and semidiurnal variation and spring-neap variation) and wind (wind set-down and wind set-up);
- The relation between river discharge and the maximum discharge during a tide, i.e. the damping of the tide by river discharge;
- The effects of a series of tides on the development of a scour hole;
- The schematised effects of one tide with a constant flow velocity;

These aspects have been studied and tested in the field verification in the Brouwersdam sluice (Figure 6.9).

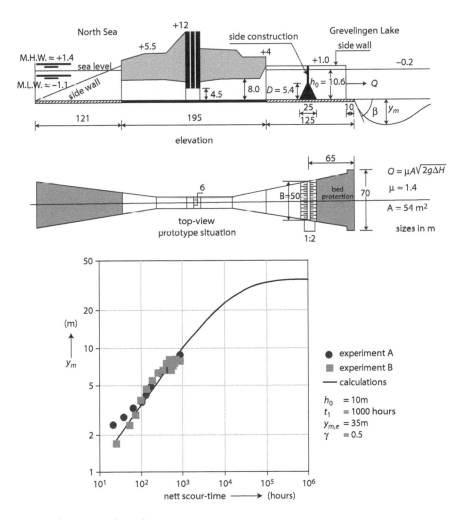

Figure 6.9 Brouwersdam sluice.

A scour hole will develop under non-steady flow conditions, and therefore, the equations for steady flow conditions are extended to unsteady flow conditions. In situations where the equilibrium phase has not been reached yet, the maximum scour depth can be given by:

$$\frac{y_m(t)}{h_0(0)} = \left(\frac{t}{t_{1,u}}\right)^{\gamma}$$ (6.13)

where
$\quad h_0(0)$ = tide-averaged flow depth (m)
$\quad t_{1,u}$ = characteristic time at which $y_m = h_0(0)$ (s)

Equation (6.8) can be adapted for unsteady flow by taking into account a succession of infinite short-lasting steady situations. If the flow is cyclic (i.e. tidal movement), the characteristic time can be represented by:

$$t_{1,u} = \frac{K h_0(0)\Delta^{1.7}}{\dfrac{1}{T}\displaystyle\int_{t_1}^{t_2}\dfrac{(\alpha U_0(t)-U_c)^{4.3}}{h_0(t)}dt}$$ (6.14)

where
$\quad T = t_2 - t_1$ half tidal period where $\alpha U_0 > U_c$ (s)
$\quad t_1$ = time at which αU_0 first exceeds U_c during flood tide (s)
$\quad t_2$ = time at which αU_0 drops below U_c during ebb tide (s)

Equations (6.13) and (6.14) have been validated for tidal conditions in the Brouwersdam prototype tests and are generally applicable. It is recommended to adjust the description of the sediment transport in Equation (6.14) in order to fulfil the Partheniades erosion law.

In Equation (6.14), the mean velocity and the flow depth are the only variables which depend on time. If $h_0(t)$ and $U_0(t)$ are given as a function of time, $t_{1,u}$ can be calculated by numerical integration. In tidal areas with a dominant tidal period of 12.4 hours (semidiurnal), a time step of 0.5 hour is recommended for the numerical integration. In deep water, the variations in $h_0(t)$ are relatively small. However, the variations in $U_0(t)$ may be quite important. As a result of ebb and flood, both the mean velocity and the sediment transport vary in time. The scour process in tidal areas can be simulated reasonably well by applying a characteristic constant mean velocity. When the mean velocity in a tidal area is approximated by a sine function, the following relation is obtained (e.g. Hoffmans, 1992):

$$U_d = \eta U_{m,t}$$ (6.15)

where
$\quad U_d$ = characteristic tidal mean flow velocity (m/s)
$\quad U_{m,t}$ = maximum velocity during a tide (m/s)
$\quad \eta$ = coefficient (–), η = 0.75–0.85

The coefficient η is almost independent of the type of sediment transport equation. For a first approximation, the characteristic time can be approximated by a simplification of Equation (6.14):

$$t_{1,u} = \frac{K h_0^2(0) \Delta^{1.7}}{(\alpha \eta U_{m,t} - U_c)^{4.3}}$$

(6.16)

6.3.7 Upstream supply of sediment

The theory in the foregoing sections has been based mainly on the results of physical scale models in which generally no upstream supply of sediment is present. However, in prototype situations, an upstream supply of sediment is often present and in such cases the development of a scour hole is reduced because part of the transport capacity of the flow is used to supply sediment from upstream and is therefore not available for the transport of sediment in the scour hole. An approximate calculation method for this reduction is discussed here. The volume unit width of a 2D scour hole representation can be expressed by the geometrical formula:

$$V(t) = c_a y_m^2(t)$$

(6.17)

where
  c_a = shape factor of scour hole (–), $c_a = 22$
  $V(t)$ = volume of scour hole per unit width (m³/m)
  $y_m(t)$ = maximum scour depth (m)

Mosonyi and Schoppmann (1968) reported that the shape factor c_a equals approximately six for clear-water scour experiments. In their experiments, the flow was nearly uniform at the end of the bed protection. When the scour hole is schematised as a triangle with an upstream slope of 1V:2H and a downstream slope of IV:8H, the shape factor c_a is 5. However, in a situation with a considerable supply of upstream sediment the upstream scour slope is about 1V:4H and the downstream scour slope 1V:40H, giving a shape factor of 22.

In the stabilisation phase, the scour hole develops towards its equilibrium. The shapes within these phases are not similar as the distance from the end of the bed protection to the cross section of maximum scour depth increases more than the maximum scour depth. Therefore, the shape factor c_a is not constant. The upstream sediment supply is defined as the volume of the sediment particles including the porosity of the sediment. The volume of the scour hole is reduced with respect to the start of the stabilisation phase by the time-dependent generally applicable formula:

$$V_r(t) = V(t) - q_s t$$

(6.18)

where
  $V_r(t)$ = reduced volume of scour hole per unit width (m³/m)
  q_s = reduction transport per unit width (including porosity) (m²/s)
  t = time (s)

Consequently, the generally applicable scouring capacity can be given by differentiating Equation (6.18):

$$\frac{dV_r}{dt} = \frac{dV}{dt} - q_s \tag{6.19}$$

The reduced maximum scour depth on prototype scale can be deduced either by combining Equations (6.17) and (6.18) or directly from Equation (6.19). As an alternative, Konter and van der Meulen (1986) give the generally applicable maximum scour depth as a function of time (Figure 6.10):

$$\frac{y_{m,p}^2(t_p + \Delta t_p) - y_{m,p}^2(t_p)}{\Delta t_p} = \frac{y_{m,c}^2(t_c + \Delta t_c) - y_{m,c}^2(t_c)}{\Delta t_c} - \frac{q_s}{c_a} \tag{6.20}$$

where
t_c = time referring to conditions where $q_s = 0$ (s)
t_p = time referring to live-bed conditions (s)
Δt = time step (s)

For the application of this method of reduction, it is necessary to determine the magnitude of q_s, which can be obtained with the aid of a scale model or by using morphological models based on sediment transport theories. For a first approximation, the unknown q_s at the end of the bed protection can be represented by:

$$q_s = \eta_b s_b + \eta_s s_s \tag{6.21}$$

where
s_b = bed load (m²/s)
s_s = suspended load (m²/s)
η_b = coefficient (–)
η_s = coefficient (–)

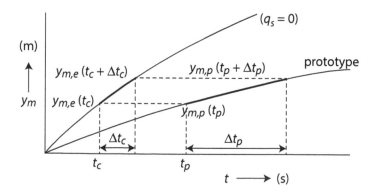

Figure 6.10 Reduction method.

Considering non-steady flow, Equation (6.21) is generally applicable, although calibration of the suspended transport is recommended in particular in tidal rivers with large water depths.

Van Rijn (1984) verified several transport predictors, using 486 sets of river data. He showed that bed load and suspended load, respectively, could be described best by:

$$s_b = 0.005\, U_0 h_0 \left(\frac{U_0 - U_c}{\sqrt{\Delta g d_{50}}} \right)^{2.4} \left(\frac{d_{50}}{h_0} \right)^{1.2} \tag{6.22}$$

$$s_s = 0.012\, U_0 h_0 \left(\frac{U_0 - U_c}{\sqrt{\Delta g d_{50}}} \right)^{2.4} \left(\frac{d_{50}}{h_0} \right) D_*^{-0.6} \tag{6.23}$$

where D_* is the sedimentological diameter (Equation 4.13).

A verification analysis of Equation (6.22) for bed load using 600 data points shows that about 77% of the predicted bed-load transport rates are within 0.5 and 2.0 times the observed values (van Rijn, 1984). As a verification for Equation (6.23) for suspended load, he used 800 data points, showing that 76% were between 0.5 and 2.0 times the observed values.

In the Eastern Scheldt, the suspended load is dominant in comparison to the bed load ($s_s \approx 10 s_b$), largely due to the relatively large flow depths (15–40 m) and the fine sediments in the deltaic area. Raaijmakers et al. (2012) refer to computational results of Stroeve in 1994 showing that the reduction method (see Section 6.6.4) yields reasonable results for the scour process in the Eastern Scheldt when the magnitude of q_s is taken as the sum of bed load and part of the suspended load ($\eta_s = 0.15$, $\eta_b = 0.85$) (Table 6.3).

Table 6.3 Hydraulic data, Eastern Scheldt (Stroeve, 1994 in: Raaijmakers et al., 2012)

	Section RO960	Section RO1680	Section HO620
Experimental data			
Length of bed protection (m)	650	650	650
Flow depth (m)	40	26	25
Sill height (m)	16.5	17.5	15.5
Mean particle diameter (μm)	200	200	200
Fluid density (kg/m^3)	1025	1025	1025
Sediment density (kg/m^3)	2650	2650	2650
Computational data			
Relative turbulence intensity (−)	0.16	0.23	0.19
Turbulence coefficient (−)	2.29	2.63	2.47
Maximum (local) velocity[a] (m/s)	1.20	1.56	1.49
Characteristic flow velocity (m/s)	0.96	1.25	1.19
Critical velocity (m/s)	0.4	0.4	0.4
Reduction transport (m^2 per day)	4.6	17.8	14.3

[a] Tidal-averaged value.

6.4 Upstream scour slopes

6.4.1 Introduction

The upstream slope of the scour hole (upstream scour slope) determines the stability of the upstream part of the scour hole and the adjacent bed protection. In general, this part of the slope reaches an equilibrium that is less steep than the initial tangent at the end of the bed protection. When this slope exceeds a critical value in non-cohesive sediments, a shear failure can occur or a liquefaction of the soil under the bed protection may even be possible. Hoffmans (1993) derived a hydraulic and morphological relation for upstream scour slopes. He calibrated this relation by using a large number of flume experiments in which the material properties and the hydraulic and geometrical conditions varied.

6.4.2 Hydraulic and morphological stability criterion

The stability of the upstream scour slope is the result of the interaction between fluid motion and material properties. The equilibrium situation of upstream scour slopes for non-cohesive material is achieved by equating the bed load due to the instantaneous bed shear stresses sloping downward and bed load due to the instantaneous bed shear stresses sloping upward. Using a probabilistic model for bed load transport, a semi-empirical relation for the slope angle β has been found (Hoffmans, 1993; Hoffmans & Pilarczyk, 1995). The equilibrium condition of bed load transport due to up-slope and down-slope shear stress resulted in a relation with a shear stress factor and a turbulence factor representing the skewness of the instantaneous bed shear stress (Figure 6.11):

$$\beta = \arcsin\left(2.9 \times 10^{-4} \frac{U^2}{\Delta g d_{50}} + (0.11 + 0.75 r_0) f_C \right) \tag{6.24}$$

where
   d_{50} = median grain size (m)
   r_0 = relative turbulence intensity at end of bed protection (–)
   U = depth-averaged velocity at end of bed protection (m/s)
   f_C = roughness coefficient (–) $f_C = C / C_0$ with $C_0 = 40\,\mathrm{m}^{0.5}/\mathrm{s}$
   C = Chézy coefficient $\mathrm{m}^{0.5}/\mathrm{s}$

Equation (6.24) is a semi-empirical equation calibrated and validated using laboratory experiments. Verification in prototype is limited to two experiments.

It appears from Equation (6.24) that turbulence is important in the development of the upstream scour slope. Steeper slopes are found when upstream conditions are more turbulent. High turbulence intensity can result in considerably steeper slopes. A smooth bed protection results in a steeper upstream scour slope too because the near-bed velocities have greater momentum and cause a more rapid expansion of the flow in the scour hole. In general, the influence of the shear stress factor can be neglected if the flow velocities are smaller than 1 m/s. In deltaic areas with fine sediments and flow velocities larger than 1 m/s, the shear stress factor determines the upstream scour slope to a large extent. For the design, it is obvious that the value of β has to be well below the critical value of the natural slope of sediment in water. Equation (6.24)

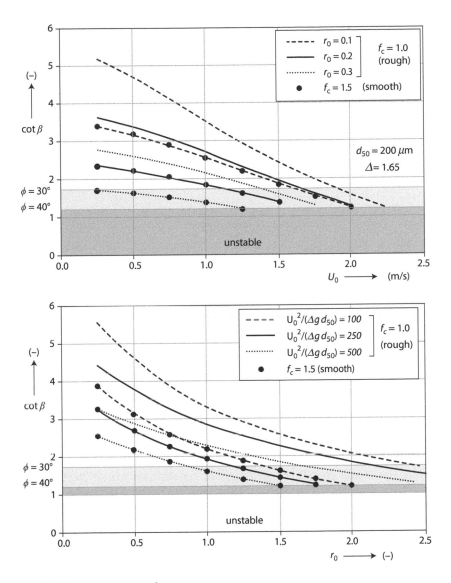

Figure 6.11 **Upstream scour slopes.**

yields results that compare reasonably well with measured developments of a scour hole in the case of a subcritical flow upstream (Figure 6.12).

6.4.3 Undermining

The gradual and dangerous undermining of the edge of the bed protection results from the turbulent energy and the erosion capacity of the flow in the recirculation zone. When β exceeds a critical value (angle of internal friction), the bed protection could be gradually undermined owing to small-scale shear failures. In addition to the gradual undermining,

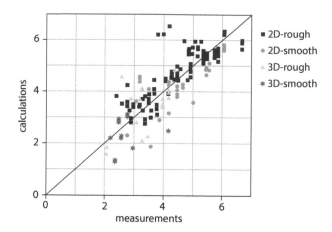

Figure 6.12 Calculated and measured cot β.

a sudden undermining may occur (shear failures and flow slides), leading to a possible failure of the hydraulic structure upstream of the scour hole. However, these phenomena are strongly dependent on geotechnical conditions (de Groot et al., 1992).

According to Konter et al. (1992), a failure length equal to eight times the maximum scour depth can be conceived as a conservative extreme length when the subsoil consists of fine and unconsolidated sand (Section 4.4). For densely packed sand, the maximum failure length can be decreased considerably and is estimated to be $L_s \approx 2y_m$. If the soil is non-homogeneous, no general rules are available, although $L_s \approx 8y_m$ can be taken as a conservative length for inhomogeneous soil profiles too.

6.5 Additional measures

Hydraulic structures placed in waterways or coastal seas are often streamlined in order to reduce flow drag, the size of wakes, and turbulence intensity. Streamlining by means of deflectors and guide vanes, however, is effective only when the hydraulic structure is aligned with the flow within narrow limits.

When the subsoil has the potential to liquefy, measures have to be taken to ensure the safety of hydraulic structures. The most obvious method is to focus the design procedure on a bed protection length such that, with a chosen bed protection, backward erosion does not influence structural stability. Local scour downstream of hydraulic structures can be reduced by lengthening and roughening the bed protection. Consequently, flow velocities and turbulence intensities reduce and the probability of flow slides and shear failures at the end of the bed protection decreases.

When the risk is too large, for example due to lack of space or for economic reasons, other measures can be considered. By protecting the upstream scour slope with rock or slag and by compacting the subsoil, shear failures and flow slides can be avoided. The construction of a retaining wall, i.e. sheet piling, at the edge of the bed protection can also be considered. It is always advisable to monitor the development of the scour hole frequently during construction and operation, so that necessary measures can be taken in time to prevent dangerous situations. During the construction phase,

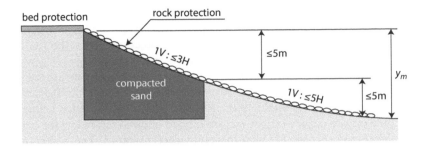

Figure 6.13 **Additional measures.**

equipment and materials are available to stabilise the upstream scour slope in order to minimise the scour that may occur under operational conditions. In the case of a land-based bed protection, it is recommended that an artificial scour hole should be made beforehand and that the upper part of the upstream scour slope should be protected by gravel or slag (Pilarczyk, 1984). In general, this will lead to a shorter and more reliable design of the bed protection. Some additional measures are illustrated in Figure 6.13.

6.6 Field experiments

6.6.1 Introduction

When sluices and dams have to be built on loosely packed sediments in deltaic areas, the scour time factor (t_1, characteristic time at which $y_m = h_0$) is important. A closure dam usually has a temporary function and equilibrium depths will not be reached for every building stage. Therefore, the scour process as a function of time has to be known since it may play an important role in the construction strategy. In such cases, we recommend applying the relations for the maximum scour depth, as given in Section 6.3.

De Graauw and Pilarczyk (1981) carried out field experiments within research regarding scour behind the storm surge barrier and compartment dams in the Eastern Scheldt. For this purpose, the sluice in the Brouwers Dam was chosen. This sluice was built to refresh the brackish water in the Grevelingen lake for environmental reasons. Aims of the experiments were to study the influence of clay layers on scour and to verify scour relations obtained from scale models.

6.6.2 Hydraulic and geotechnical conditions

The discharges and flow velocities encountered during two field experiments A and B at the sluice in the Brouwersdam were almost identical, whereas the soil characteristics were different. The discharges, the flow levels, and the bed configuration were measured frequently and some flow velocity and concentration measurements in the centre of the sluice were also carried out. During the experiments, the sea water was let into the lake during the flood and was released during the ebb. The outflow had no influence on the development of the scour hole because of the relatively small flow velocities above the scour hole during the ebb, while the suspended load transported from the sea into the lake was also negligible.

A 5.4m high sill was constructed on the lake side of the sluice with two side constrictions measuring 2.5 m on the left side and 1.5 m on the right side. The flow depth was about 10 m and the length of the bed protection from the toe of the sill measured about 50 m. The effective roughness of the bed protection was estimated to be 0.4 m. The other dimensions of the sluice are presented in Figure 6.9.

The soil characteristics with respect to Experiment A were measured beforehand. The diameter of the bed material varied with the depth from 0.2 to 0.3 mm. Some thin clay lenses were present, especially in the soil layer between 2 and 4 m below the original bed. The thickest clay layer of 0.2 m was situated at about 3.5 m below the bed. The scour hole that developed was refilled with loosely packed material. The bed material regarding Experiment B consisted of fine sand with a particle diameter of about $d_{50} = 0.26$ mm. The particle diameter for which 90% of the mixture is smaller than d_{90} measured 0.29 mm. As a result of tidal variations, both the flow velocity and the sediment transport varied. To simulate the scour process a theoretical characteristic mean velocity was introduced, which was defined as the mean velocity; this would give the same average sediment transport.

6.6.3 Discussion

Figure 6.14 shows some measured bed profiles of the prototype experiments at different times. The gradual undermining, including a shear failure, is shown in Figure 6.15, and Figure 6.9 shows the maximum scour depth as a function of time.

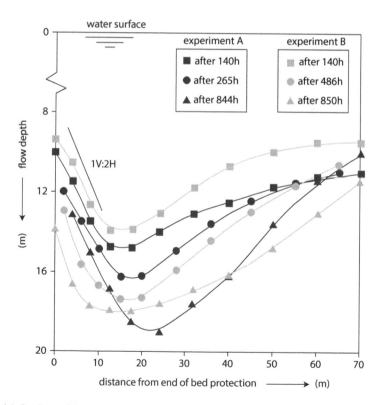

Figure 6.14 Bed profiles of scour holes (Brouwersdam).

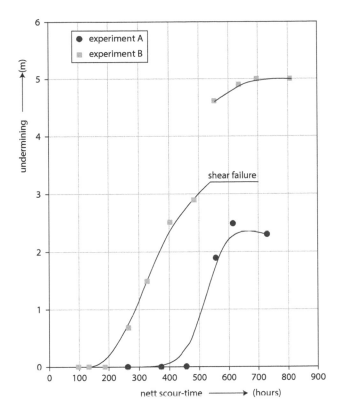

Figure 6.15 Undermining as a function of time (Brouwersdam).

6.6.3.1 Upstream scour slope

Due to the tidal influence, the flow velocities varied in time. In experiments A and B, the maximum velocities averaged over 140 tides were approximately 1.2 m/s (Hoffmans, 1992). From this velocity and using Equation (6.24), it follows that cot $\beta \approx 1.8$ or 1V:1.8H. More details of experimental and computational results can be found in Tables 6.4 and 6.5 and in Delft Hydraulics (1979).

With respect to the upstream scour slope, it was observed that the variation in the slope over the first few metres was rather large. Due to the flow velocities and the high turbulence generated in the mixing layer and in the vortex street, the slope became steeper until the critical slope of about 1V:1.5H was reached. Then, a shear failure (Experiment B) occurred, resulting in a milder slope for the upstream scour slope. After this instability, the steepening of the upstream scour slope started again. At the end of the prototype experiments, the upstream scour slope measured approximately 1V:2H.

6.6.3.2 Undermining

In general, the angle of internal friction for sand lies in the range of 30–40°, depending on the porosity, the particle diameter and the grain size distribution of the mixture. Applying Equation (6.24), small-scale shear failures can be expected when the flow

Table 6.4 Experimental results of field experiments at Brouwers Dam

Experimental parameters	Experiment A	Experiment B
Initial flow depth (scour hole) (m)	10.6	9.6
Height of sill (m)	5.4	5.4
Length of bed protection (m)	50	52
Effective roughness of bed protection (m)	0.4	0.4
Averaged discharge (m³/s)	271	270
Maximum discharge (m³/s)	380	380
Particle diameter (mm)	$d_{50} = 0.25$ $d_{90} = 0.29$	$d_{50} = 0.26$ $d_{90} = 0.29$
Upstream scour slope at end of test	IV:2.2H	IV:1.5H
Undermining just before shear failure slide (m)		2.9
Undermining at end of test (m)	2.3	5.0
Condition subsoil	Clay/sand	Sand
Characteristic time (hours) (extrapolated)	2000	800

Table 6.5 Computed values for the field experiments at Brouwers Dam

Computational parameters	Experiment A	Experiment B
Characteristic discharge (m²/s)	9.89	9.89
Characteristic mean velocity (m/s)	0.93	1.03
Critical mean velocity (m/s)	0.41	0.41
Relative turbulence intensity (−)	0.28	0.29
Turbulence coefficient (−)	2.91	2.92
Roughness function (−)	1.13	1.12
Upstream scour slope	IV:1.8H	IV:1.8H
Characteristic time (hours)	2400	1200
Equilibrium scour depth (m)	33.6	35.5

velocities in the prototype situation are larger than 1.3 m/s for ($\phi' = 30°$) or 1.9 m/s for ($\phi' = 40°$). During the experiments, flow velocities were measured varying from 1.5 to 2.0 m/s. Since both small-scale shear failures and a large shear failure that occurred after approximately 450 hours (Experiment B) were observed, Equation (6.24) seems to be feasible for use in practical engineering. When the subsoil consists of clay and sand layers the results obtained from Equation (6.24) must be interpreted carefully because the influence of the cohesion of the subsoil has not been taken into account.

6.6.3.3 Time scale

Both experiments were ended after about 800 hours net scour time. At that time, the maximum scour depth in both experiments was approximately equal to the initial flow depth. The characteristic time at which the maximum scour depth equals the initial flow depth was extrapolated from the measurements when no shear failures would occur. These measured times were compared with the computed ones obtained with Equation (6.17). The discrepancy ratio r (ratio between the calculated and measured time scale) for both experiments lies in the range of $2/3 < r < 1.5$ (Tables 6.4 and 6.5). Note that the degree of turbulence and the flow velocities determine the scour time scale to a large extent. Hoffmans (1992) showed that r is larger than two (or smaller than 0.5) if the error in the discharge is greater than 15%. Thus, the computed and observed time scales do not show a large difference.

6.6.3.4 Equilibrium scour depth

Besides the upstream scour slope, the maximum scour depth in the equilibrium phase also determines the optimal length of the bed protection downstream of hydraulic structures. The calculated equilibrium scour depth is about three times the initial flow depth. The time required to reach this depth in the prototype situation is extremely long. After about 25 years net scour time, the equilibrium scour depth will almost be achieved, as can be seen in Figure 6.9.

6.6.3.5 Evaluation brouwers dam experiments

The objective of these field experiments was to verify the scour relations obtained from scale models (Breusers, 1966, 1967; van der Meulen & Vinjé, 1975; de Graauw & Pilarczyk, 1981). The computed results compare favourably with the measured ones, so the hydraulic and morphological relations seem to be applicable for practical engineering. However, the scour process is not only influenced by turbulence parameters but also by the geotechnical ones. Currently, the instabilities of the subsoil, especially the phenomena of flow slides, are being researched extensively. To produce a safe and reliable design, the total reliability of all modes of failure should be approximated at least to a conceptual level. A fault tree is a useful tool for integrating the various mechanisms into a single approach.

6.6.4 Experiences Eastern Scheldt

With a length of 9 km, the Eastern Scheldt storm surge barrier is one of the largest flood defenses in the world (Figure 6.16). On each side of this barrier, a scour protection with a length of 550 650 m prevents scour from endangering its stability. Van Velzen et al. (2012, 2014) describe the scour development downstream of the bed protection based

Figure 6.16 The Eastern Scheldt barrier, consisting of the 'Hammen' in the north (top left corner), the 'Schaar' in the centre and the 'Roompot' inlet in the south (bottom right corner).

on an extensive bathymetrical data set for the years 1985–2012. This section describes scour results in the Roompot-East channel. The measured scour depths are discussed in relation to the geotechnical events until 2012. It also discusses the range of scour depths predicted for 2050.

Figure 6.17 shows the formation of two scour holes in 'Roompot-East' and demonstrates that the maximum water depth in both scour holes is about −59 m + NAP in 2012. When relating the scour depth to the initial height of the edge beam at the end of the bed protection, the actual scour depth is obtained; the northern scour hole is approximately 34 m deep, whereas the southern scour hole is 21 m deep.

As the two scour holes at 'Roompot-East' deepened and widened, both the upstream slope and the side slopes of the scour holes became so steep that geotechnical instability events occurred perpendicular and parallel to the barrier axis, respectively. In total, 12 geotechnical instability events were identified at 'Roompot-East' in the period 1985–2012. More information can be found in Stoutjesdijk et al. (2012).

In 1993, a loose rock (slags) protection was applied on a large part of the upstream slope, which stabilised the upstream slope. However, the scour hole in the northern part deepened further and steep slopes could develop below the protected slope, which eventually led to some geotechnical instability events perpendicular to the barrier axis along the upstream slope in the period 2002–2005 (Figure 6.18).

After the instability of 2003, the upstream slope became even steeper. The maximum steepness of the upstream slope over a height of 5 m was 1V:3H in 2002 and 1V:2H in 2003. In 2008, a large geotechnical instability event occurred along the upstream slope, which did not only undermine the maintenance works but also lowered the edge beam by about 5 m.

Figure 6.18 furthermore shows that a clay layer was reached in 2006, which has been eroding locally from 2011 onwards. The scour development was only temporarily slowed down due to the presence of a clay layer. It has not yet reached equilibrium. Based on the reduction method, some computational results are presented.

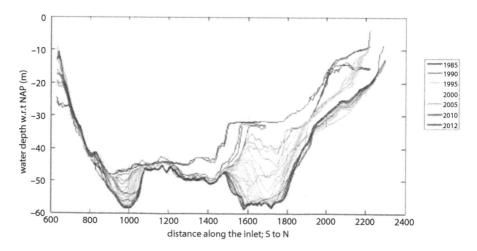

Figure 6.17 Maximum water depth along the barrier at 'Roompot-East' with respect to NAP.

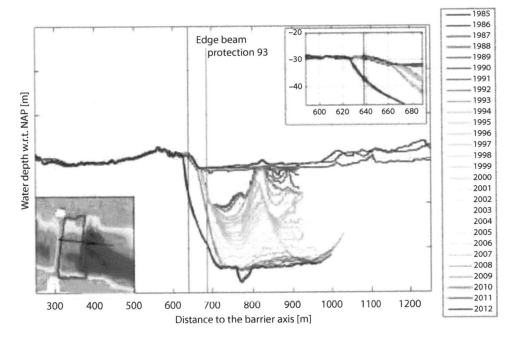

Figure 6.18 Cross-section of the bathymetry perpendicular to the axis of the barrier at 'Roompot-East'.

In Raaijmakers et al. (2012), refer is made to a prediction by Stroeve in 1996 that in 2050 a water depth of 62 m at Roompot-East might be expected. Because of a deep clay layer of some metres thick, Van Velzen et al. (2012, 2014) predicted that the lower and upper limits of the water depth would measure 60 m and 125 m, respectively. Hence, the influence of the thickness on the scour process is significant; see also Figure 6.19 which gives the scour depth as a function of time. Clearly, the difference can be seen between a conservative approach and the probable scour. Also, the effect of clay layers is evident. At a depth of 30 m, the clay layer decreases the rate of scouring.

6.7 Example

6.7.1 Introduction

This example addresses the computation of the critical upstream scour slope as a function of the flow velocity and the scour depth in the equilibrium phase. The results are compared with observed values in field experiments.

6.7.2 Critical upstream scour slope downstream a sill

A storm surge barrier is constructed in a wide estuary. The height of the broad-crested sill is 10 m, and the distance between the piers is 50 m. During operation, when the gates between the piers are raised to the upper state, the flow far downstream the sill is subcritical. If the water level at sea exceeds a critical value which is, on average, three times a year,

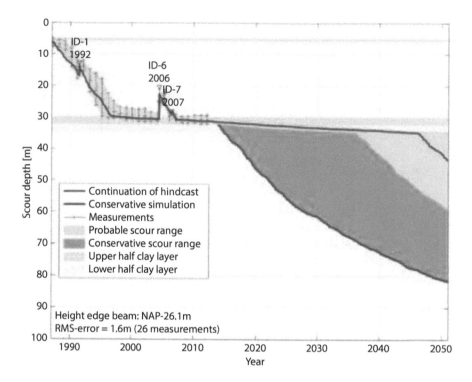

Figure 6.19 Simulation of the scour development with the reduction method at 'Roompot-East' with respect to the initial height of the edge beam (beam is about 10 m – NAP).

the barrier is closed. The flow depth in the estuary is about 25 m and the mean tidal range is 2.5 m, with maximum velocities of 1.5 m/s. During spring tide, flow velocities of 2.0 m/s are observed. The tidal curve is simplified to a sine function. The length of the bed protection is $L = 500$ m, and the effective roughness of the bed protection is estimated to be $k_s = 0.5$ m. The non-cohesive bed material characteristics are $d_{35} = 180$ µm, $d_{50} = 200$ µm, and $d_{90} = 300$ µm. The angle of internal friction is $\phi' = 35°$. The average water temperature is 15°C, $v = 1.1 \times 10^{-6}$ m²/s. Other data are $\rho_s = 2650$ kg/m³, $\rho = 1025$ kg/m³.

a. What is the characteristic mean velocity which is defined as the flow velocity that will result in the same average transport?
b. What is the relative turbulence intensity at the end of the bed protection?
c. What is α according to the methods of Jorissen and Vrijling (1989) and Hoffmans (1993)?
d. What is α if the length of the bed protection is $L = 125$ m?
e. What is the characteristic time at which the maximum scour depth equals the initial flow depth?
f. What is the slope angle of the scour hole in the equilibrium phase? Will the bed protection be undermined when no additional measures (to prevent dangerous situations) are taken in time?

g. What is the maximum scour depth in the equilibrium phase according to Dietz (1969) and motivate why this value is too large.

h. Is the length of the bed protection sufficiently long to ensure the safety of the storm surge barrier against the occurrence of flow slides, if computational results of the reduction method have shown that the maximum scour depth in the equilibrium phase is approximately 55 m?

Solution:

a. Characteristic mean velocity
Equation 6.15:

$$U_d = 0.8 \times U_{m,t} = 0.8 \times 1.5 = 1.2 \text{ m/s}$$

b. Relative turbulence intensity (at end of bed protection)
Chézy coefficient:

$$C = \sqrt{g}/\kappa \times \ln(12 \times h_0/k_s) = \sqrt{9.81}/0.4 \times \ln(12 \times 25/0.5) = 50 \text{ m}^{1/2}/\text{s}$$

Equation 6.10:

$$
\begin{aligned}
r_0 &= \sqrt{\left[0.0225 \times (1 - D/h_0)^{-2} \times \left[(L - 6 \times D)/(6.67 \times h_0) + 1 \right]^{-1.08} + 1.45 \times g/C^2 \right]} \\
&= \sqrt{\left[0.0225 \times (1 - 10/25)^{-2} \times \left[(500 - 6 \times 10)/(6.67 \times 25) + 1 \right]^{-1.08} + 1.45 \times 9.8/50^2 \right]} \\
&= \sqrt{\left[0.0155 + 0.0057 \right]} = 0.15
\end{aligned}
$$

c. Flow and turbulence coefficient α ($L = 500$ m)
Jorissen & Vrijling:

$$\alpha = \alpha_l = 1.5 + 5 \times r_0 = 1.5 + 5 \times 0.15 = 2.25$$

Hoffmans:

$$\alpha = \alpha_l = 1.5 + 4.4 \times r_0 \times C/C_0 = 1.5 + 4.4 \times 0.15 \times 50/40 = 2.33$$

If no sill is present, the relative turbulence intensity is about $r_0 = 0.10$ for hydraulically rough flow conditions. Hence, the minimum value of α is 2.0 (Figure 6.8). The computation will be proceeded with $\alpha = 2.3$.

d. Flow and turbulence coefficient α ($L = 125$ m)
If the length of the bed protection is $L = 125$ m, the influence of turbulence e.g. the Karman vortex street cannot be neglected at the end of the bed protection, ratio $L/h_0 = 5$. In addition, owing to the wake zone behind the piers, the depth-averaged velocities are not constant at the transition of the fixed bed to the erodible bed. Therefore, a scour hole with a strongly 3D character will develop.

de Graauw & Pilarczyk:

$$\alpha = 6.0 \text{ (Figure 6.8)}$$

Equation (6.10):

$$r_0 = \sqrt{\left[0.0225 \times (1 - D/h_0)^{-2} \times \left[(L - 6 \times D)/(6.67 h_0) + 1\right]^{-1.08} + 1.45 \times g/C^2\right]}$$

$$= \sqrt{\left[0.0225 \times (1 - 10/25)^{-2} \times \left[(125 - 6 \times 10)/(6.67 \times 25) + 1\right]^{-1.08} + 1.45 \times 9.8/50^2\right]}$$

$$= \sqrt{\left[0.0438 + 0.0057\right]} = 0.22$$

Jorissen and Vrijling:

$$\alpha_l = 1.5 + 5 \times r_0 = 1.5 + 5 \times 0.22 = 2.60$$

Hoffmans:

$$\alpha_l = 1.5 + 4.4 \times r_0 \times C/C_0 = 1.5 + 4.4 \times 0.22 \times 50/40 = 2.71$$
$$\alpha = \alpha_l \times U_l / U_0$$

Following the method of de Graauw and Pilarczyk (1981), a conservative value of $\alpha = 6$ that corresponds to a location where the depth-averaged velocity is about at its maximum is computed. If the latter two methods are used, it is necessary to know the local depth-averaged velocities at the end of the bed protection!

e.   Characteristic time $t_{1,u}$ at which $y_m = h_0$ with $L = 500\,\text{m}$
     van Rijn:

$$D_* = d_{50} \times (\Delta \times g/\nu^2)^{1/3} = 200 \times 10^{-6} \times [1.59 \times 9.8/(1.1 \times 10^{-6})^2]^{1/3} = 4.69$$
$$\Psi_c = 0.14 \times D_*^{-0.64} = 0.14 \times 4.69^{-0.64} = 0.052 \, (4 < D_* < 10)$$
$$U_c = \sqrt{(\Psi_c \times \Delta \times g \times d_{50})}/\kappa \times \ln\left[12 \times h_0/(3 \times d_{90})\right]$$
$$= \sqrt{(0.052 \times 1.59 \times 9.8 \times 200 \times 10^{-6})}/0.4 \times \ln\left[12 \times 25/(3 \times 300 \times 10^{-6})\right]$$
$$= 0.40 \text{ m/s}$$

Equation (6.8):

$$t_{1,u} = 330 \times h_0^2 \times \Delta^{1.7}/\left[\alpha U_d - U_c\right]^{4.3} = 330 \times 25^2 \times 1.59^{1.7}/\left[2.3 \times 1.2 - 0.40\right]^{4.3}$$
$$= 11300 \text{ hours}\left(\text{or } 1.3 \text{ year}\right) \text{ net scour time}$$

Breusers' method (Equation 6.7) is based on model tests with no upstream supply of sediment and has been verified by some field experiments. In this example, sediment particles are in suspension upstream of the scour hole, so the erosion capacity of the flow is not fully used for picking up bed particles in the scour hole. This results in a slower development of the scour hole. Therefore, t_1 is larger than 11,300 hours. Values computed with Equation (6.8) must be considered as first estimations, especially when the bed consists of fine sediments.

f.   Upstream scour slope

Equation (6.24):

$$\cot \beta = \cot\left[\arcsin\left[2.9\times10^{-4}\times(U_0)^2\big/(\Delta\times g\times d_{50})+(0.11\ +\ 0.75\times r_0)\times C/C_0\right]\right]$$
$$= \cot\left[\arcsin\left[2.9\times10^{-4}\times1.5^2\big/(1.59\times9.8\times200\times10^{-6})+(0.11+0.75\times0.15)\times50/40\right]\right]$$
$$= 1.8\,(U_0 = 1.5\ \text{m/s})$$
$$\cot \beta = \cot\left[\arcsin\left[2.9\times10^{-4}\times2.0^2\big/(1.59\times9.8\times200\times10^{-6})+(0.11+0.75\times0.15)\times50/40\right]\right]$$
$$= 1.2\,(U_0 = 2.0\ \text{m/s})$$

During spring tide, the flow velocities are about 2.0 m/s. With these velocities, the computed upstream scour slope is steeper than the natural slope of the scour hole ($\cot 35° = 1.4$). Therefore, small-scale failures could be expected, so the end of the bed protection has to be protected by rock or gravel to ensure the safety of the hydraulic structure. In addition to the gradual undermining, shear failures and flow slides may occur, leading to a possible failure of the hydraulic structure. However, these phenomena are strongly dependent on geotechnical conditions. Verification by some field experiments (Brouwers Dam) showed that the predictability of Equation (6.24) is reasonable and therefore feasible for practical engineering.

g.   Maximum scour depth in the equilibrium phase

Dietz (1969):

$$y_{m,\,e} = h_0 \times (1 + 3r_0) \times U_d - U_c \big/ U_c = 25 \times [(1 + 3 \times 0.15) \times 1.2 - 0.41]/0.41 = 81\ \text{m}$$

The maximum scour depth according to Dietz is a conservative value because the influence of upstream supply of sediment is not taken into account. In deltaic areas with fine sediments, the reduction method is recommended.

h.   Geotechnical stability

The length of the bed protection is $L = 500$ m, and the maximum scour depth in the equilibrium phase is about $y_{m,e} = 55$ m. Applying the storage models of Silvis, the length of the expected damage of the bed protection is $L_s = 8.0 \times y_{m,e} = 8 \times 55 = 440$ m, provided that a flow slide with an extreme probability occurs in the equilibrium phase of the scour process. Consequently, the length of the bed protection is sufficiently long ($L > L_s$) during the development of the scour hole.

Chapter 7

Abutments and groynes

7.1 Introduction

Abutments are part of the valley side against which dams are constructed or approach embankments in the case of bridges. Groynes in rivers are usually designed to protect the banks or to provide enough flow depth for navigation purposes. Generally, groynes appear as a series of transverse structures whereas abutments are single structures. Alternative names for these structures include spurs and transverse dikes. They can be classified according to the type, for example, T-headed or L-headed abutments (guide bunds) and groynes. In addition, they can be classified according to permeability (impermeable or permeable) and to the height of the abutments and groynes below high water (submerged or nonsubmerged). Groynes and abutments can be placed on one bank or symmetrically on both banks of a waterway. Often, in practice, the design of an abutment includes also a sill. Abutments and groynes can be protected by an adjacent bed protection to prevent the formation of a scour hole in the direct vicinity of the groyne or abutment. To estimate local scour downstream of submerged groynes and abutments, the method in Chapter 6 for scour downstream of sills can be used.

Abutments and groynes are obstacles placed in a flow, in such a way that they result in a horizontal constriction with a three-dimensional flow. Bridge piers (Chapter 8) are also horizontal constrictions, but in their case, the water can flow at both sides of the structures. Unlike sills which produce vertical constrictions (Chapter 6), groynes and abutments result in horizontal constrictions mainly in a three-dimensional flow. Some characteristics of groynes and abutments and the corresponding flow pattern will be discussed in Section 7.2, after which calculation methods will be treated in Sections 7.3–7.5. Section 7.6 pays attention to failure mechanisms and measures to mitigate scour. As the main difference between groynes and abutments is merely their number, this chapter will only use the term abutments. Finally, two examples are presented in Section 7.7.

7.2 Geometry characteristics and flow patterns

7.2.1 Introduction

In general, a distinction can be made between streamlined and blunt or sharp-nosed abutments. The geometry of abutments in rivers or estuaries can be schematised to define some basic types of geometries: wing-wall, spill-through abutments and vertical-wall abutments (Figure 7.1).

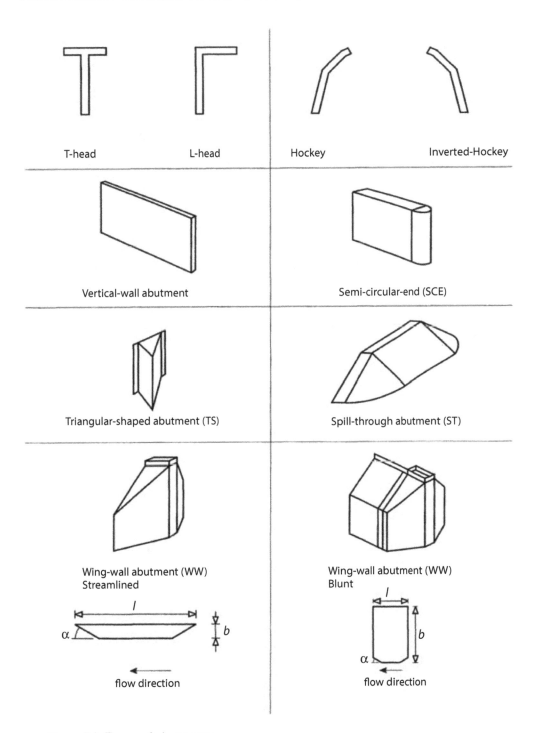

T-head   L-head   Hockey   Inverted-Hockey

Vertical-wall abutment   Semi-circular-end (SCE)

Triangular-shaped abutment (TS)   Spill-through abutment (ST)

Wing-wall abutment (WW)
Streamlined

Wing-wall abutment (WW)
Blunt

flow direction   flow direction

Figure 7.1 **Types of abutments.**

Different types of abutments produce different associated scour holes due to the different flow fields (see Section 7.2.5):

- Abutments with a bend (positively bended) show a scour hole at the bend;
- Abutments with a sudden crest lowering at the root will show an extra scour hole at that location. This type of scour hole can also be the result of irregular plants or trees on the abutments;
- At submerged abutments, scour holes will occur downstream of the abutment and are comparable with scour by 2D-V plunging jets (Chapter 5) or scour behind sills (Chapter 6);
- At abutments with an oblique flow over the crest, scouring occurs due to the so-called screw eddy downstream of the overflowing abutment;
- In an estuary, the flow direction changes with the tide, and fresh water may flow over salt water in a stratified system and creating a mixing layer of fresh and salt water. Abutments in the Nieuwe Waterweg show two scour holes: one due to flood flow and one due to ebb flow.

7.2.2 Wing-wall abutments

Wing-wall abutments have several complicated shapes, including straight wing-wall abutments with a T-head or L-head (guide bunds), curved wing-wall abutments shaped like a hockey stick or an inverted hockey stick. A wing-wall abutment along the bank of a canal, a river or a tidal channel can be depicted by the following geometric parameters (Figure 7.2): length (l), width (b), angle of the upstream wing-wall (α_1) and the angle of the downstream wing-wall (α_2). If the ratio b/l is larger than 1 and if the angles α_1 and α_2 are larger than 20°, these structures can be considered as blunt.

7.2.3 Spill-through abutments

The geometry and the shape of spill-through abutments can be characterised by the geometric parameters. Besides these parameters, the angle of the side slope (β_1) and the angle of the slope (β_2) at the nose of the abutment are also representative (Figure 7.3). The crests of abutments in a river have a gentle slope of 1:100 to 1:200 along the abutment axis to divert the overflowing flow from the bank to the centre of the flow. However, in an estuary horizontal crests are often used. The shape of the edges between the side wall and the wing-wall may be sharp or rounded (Figure 7.3). A special case is an abutment with a completely rounded head. If the ratio b/l is smaller than 0.2, the structure can be schematised as a streamlined abutment. If the angles β_1 and β_2 are smaller than 6° the local constriction scour will dominate the scour due to the vortex street downstream of the separation point.

Scour along the streambank and at the abutment tip is also influenced by the permeability of the abutment. Impermeable abutments, in particular, can create erosion of the streambank at the abutment root. This can occur if the crests of impermeable abutments are lower than the height of the bank. Under submerged conditions, flow passes over the crest of the abutment generally perpendicular to the abutment. Laboratory

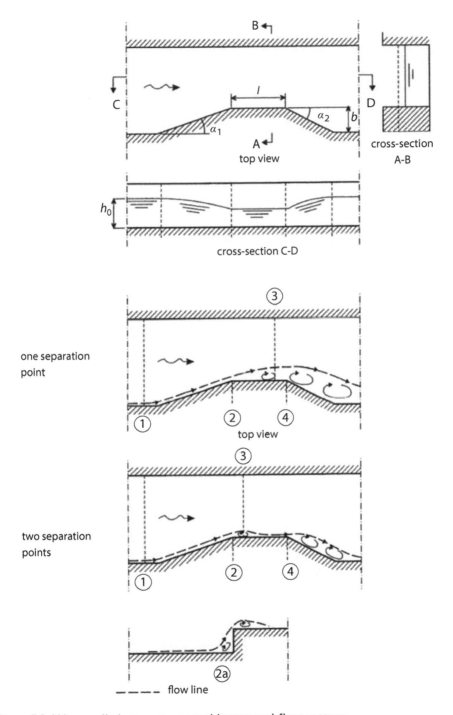

Figure 7.2 Wing-wall abutment, general layout and flow pattern.

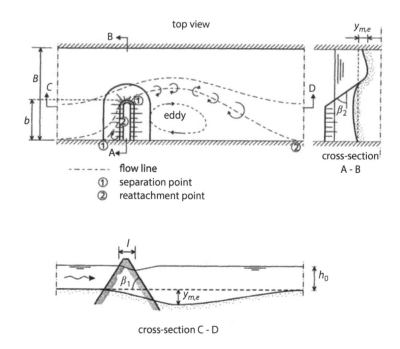

Figure 7.3 Spill-through abutment general layout and flow pattern.

studies of abutments with more than 70% permeability were observed to cause very little bank erosion, whereas abutments with permeability of 35% or less caused bank erosion similar to the effect of impermeable abutments. An example of a permeable abutment is shown in Figure 7.4.

7.2.4 Vertical-wall abutments

In scale models, a vertical-wall abutment is sometimes schematised as a vertical plate with a vertical head. In the experiments, the thickness (l) of the plate is generally small compared to the width (b) of the abutment normal to the flow direction (say $b/l > 5$). Such geometries can be considered as sharp-nosed, therefore these structures will be found in the prototype only in exceptional cases.

7.2.5 Flow pattern

The interaction between wake vortices is important for local scour near abutments and bridge piers. Carstens (1976) distinguishes three different types of interaction which are elucidated in Table 7.1. The flow field around abutments is generally characterised by an acceleration from upstream to the most contracted cross-section somewhere at or just downstream of the head of the abutment, followed by a deceleration of the flow. Downstream of the abutment the main flow is separated from a large eddy by a vortex street. Moreover, there is also an eddy with a horizontal axis. Far downstream of the reattachment point, uniform flow will be re-established.

Figure 7.4 Left: pile screen as permeable abutment (Jamuna River at Kamarjani, Bangladesh). Right: schematised permeable abutment.

Table 7.1 **Flow characteristics around hydraulic structures**

Geometry	Interaction between geometry and flow
$b/h_0 < 0.5$	A strong interaction, vortices are generated intermittently from a separation point at the left and the right sides of the structure
$0.5 < b/h_0 < 1.5$	A weak interaction
$b/h_0 > 1.5$	No interaction, the vortices are generated independently from a point at the left and the right sides of a structure

A separation point and a small eddy may occur just upstream of the abutment (Figure 7.2). If the angle α_1 is near 90°, a separation point occurs with small vortices in the corner between the bank and the abutment. Near this corner, a surface roller may be generated. The down flow at the vertical-wall abutment can generate a strong spiral motion near the bed. The flow will reattach at the upstream side of the abutment if the length parallel to the main flow is sufficiently long. In such cases, a second separation point will occur (Figure 7.2). These points are the start of vortex streets that can cause serious scour holes in the bed. However, also the eddies with a horizontal axis cause scour. A rather complex three-dimensional flow pattern appears if an abutment is submerged, sometimes with a hydraulic jump just downstream of the abutment.

Abutments projecting into wide floodplains may produce scour in two ways. First, a strong concentration of streamlines and a corresponding increase of flow velocities can produce deep scour holes upstream of the contraction. In many cases, this results in a serious scour potential at the abutment. Second, the strong vortices next to the dead water region produces the maximum scour depth upstream of the abutment (Figure 7.5). Upstream of Chandpur in the lower Meghna river, scour depths of more than 70 m were observed, these being comparable to scour depths downstream of the contraction (Haskoning, 2002).

This chapter addresses local scour at hydraulic structures for which the ratio between the width (b) and the flow depth (h_0) is larger than 1. Calculation methods for geometries where $b/h_0 < 1$, for example scour at slender bridge piers, are discussed in Chapter 8. It should be noted that some methods can be applied for all the three types of flow, as shown in Table 7.1 (e.g. the Breusers method).

Klaassen et al. (2012) describe scour at guide bunds; see Figure 7.6, showing in a topview the guide bund with the projected flow lines, large eddies and vortex streets as

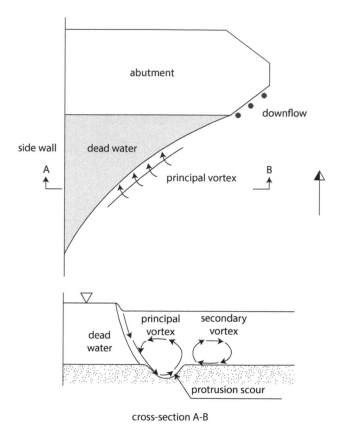

Figure 7.5 **Flow features near an abutment.**

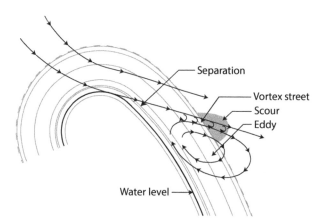

Figure 7.6 **Example of scour at a guide bund of the Jamuna Bridge in Bangladesh due to changed flow pattern during outflanking.**

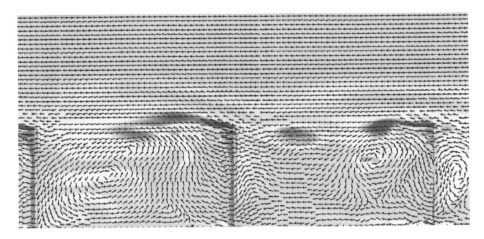

Figure 7.7 **Example of the flow through a permeable abutment.**

a result of the different approach flow directions due to upstream channel migration. The vortex streets generate high turbulence levels responsible for the scour holes.

The recently constructed longitudinal dams in the Dutch rivers are in a way comparable with guide bunds; however, they lack the characteristic curls and their shape and intended functioning are different. At least scour holes can be expected at the head and at the end of these dams and also at the sides depending on the angle of the flow attack. Obviously, the flow field at a permeable abutment is quite different, as shown in Figure 7.7.

7.3 Dutch modelling

7.3.1 Introduction

The development of local scour around abutments can be divided into an initial phase, a development phase, a stabilisation phase and an equilibrium phase (Section 3.4.1). In general, local scour around hydraulic structures develops relatively rapidly. For engineering purposes, a time-dependent description of the maximum scour depth is not always relevant. However, in deep rivers, the time factor may be important, for example during the construction phase.

7.3.2 Breusers approach

For abutments ($b/h_0 > 1$), the maximum scour depth can be described by the following relation, which is valid for all phases of scour development, provided $y_{m,e} > h_0$ (also Equation 3.6):

$$\frac{y_m}{y_{m,e}} = 1 - e^{-\frac{h_0}{y_{m,e}}\left(\frac{t}{t_1}\right)^\gamma} \tag{7.1}$$

where

h_0 = initial flow depth (m)
t = time (s)
t_1 = characteristic time (s) at which $y_m = h_0$ as long as $y_{m,e} > h_0$
y_m = maximum scour depth at t (m)
$y_{m,e}$ = maximum scour depth in the equilibrium phase (m)
γ = coefficient (–), $\gamma = 0.4$

The time-dependent growth of the maximum depth of a scour hole in the development phase, that is for $t < t_1$, can be given by the generally applicable relation of Breusers (1966) (also Equation 4.7):

$$\frac{y_m}{h_0} = \left(\frac{t}{t_1}\right)^{\gamma} \tag{7.2}$$

From a dimensional analysis and many experiments, the following relation for t_1 could be deduced:

$$t_1 = \frac{K h_0^2 \Delta^{1.7}}{(\alpha U_0 - U_c)^{4.3}} \tag{7.3}$$

where

K = coefficient (if $K = 330\ m^{2.3}/s^{4.3}$ then t_1 is expressed in hours)
U_c = critical mean velocity (m/s)
$U_0 = Q/A$, mean velocity (m/s)
Q is discharge (m^3/s)
A is cross section (m^2)
α = coefficient depending on the flow velocity and turbulence intensity (–)
Δ = relative density (–)

The exponent 4.3 in Equation (7.3) was confirmed by a scale model study of the piers for the Eastern Scheldt Storm Surge Barrier (Akkerman, 1976). The coefficients α and K follow from model test results (Table 7.2). However, note that the characteristic time scale t_1 is not dimensionless; see also the example in Section 7.7.

As mentioned in Chapter 3, the α factor depends on flow velocity and turbulence intensity. This concept could also be applied for other types of structures, such as abutments and bridge piers. For abutments, this coefficient is strongly related to the contraction ratio m (i.e. ratio between width of the structure and width of the river).

Table 7.2 Scour coefficients

Source	Geometry	b/h_0	α	K
Akkerman (1976)	Elongated pier	1.2	7.15	250
Konter (1982)	Cylindrical pier	1.2–1.9	6.0–8.0	330
van der Wal (1991)	Vertical wall abutments	1.0–7.5	4.0–9.0	330
van der Wal (1991)	Spill-through abutments	1.0–3.0	2.0–7.0	330

7.3.3 Closure procedures

In the foregoing, attention has been paid to permanent hydraulic structures. However, structures can also be temporary, for example those used during the closure of a river branch or a tidal channel. In principle, three methods for the closure of a waterway can be distinguished:

1. Vertical closure,
2. Horizontal closure,
3. A combination of both vertical and horizontal methods of closure.

A gradual vertical constriction is obtained by increasing the sill height during the construction period. In such cases, the coefficient α can be estimated by using equations for two-dimensional scour since the turbulence intensity is constant in the transverse direction. For a horizontal closure where the relative turbulence intensities and flow velocities are not constant along the width of the flow, α can be obtained from Figure 6.8. However, prudence is called for when dealing with complex hydraulic structures.

A vertical closure with abutments or a horizontal closure generates a flow pattern with two vortex streets downstream of the separation point on the head of the horizontal constriction (Figure 7.8). The horizontal position of the vortex street, which depends on the flow field, determines where along the downward edge of the bed protection the highest values of α can be expected. When the sill height D is relatively low $(D/h_0 < 0.3)$, the angle between the vortex street and the main flow direction is smaller than 45° (1:1), whereas for high sills $(0.3 < D/h_0 < 0.6)$, the angle lies in the range of 45° (1:1) to 70° (1:2).

Scour due to three-dimensional flow can be predicted by applying the Breusers equilibrium method (Equations 7.1 and 7.2) in combination with local flow velocities and turbulence intensities (Figure 7.9). These values can be obtained from mathematical

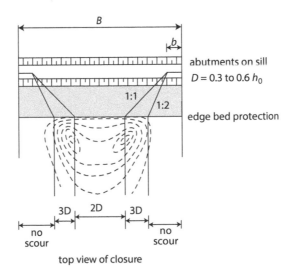

Figure 7.8 **Type of scour hole as function of the river width.**

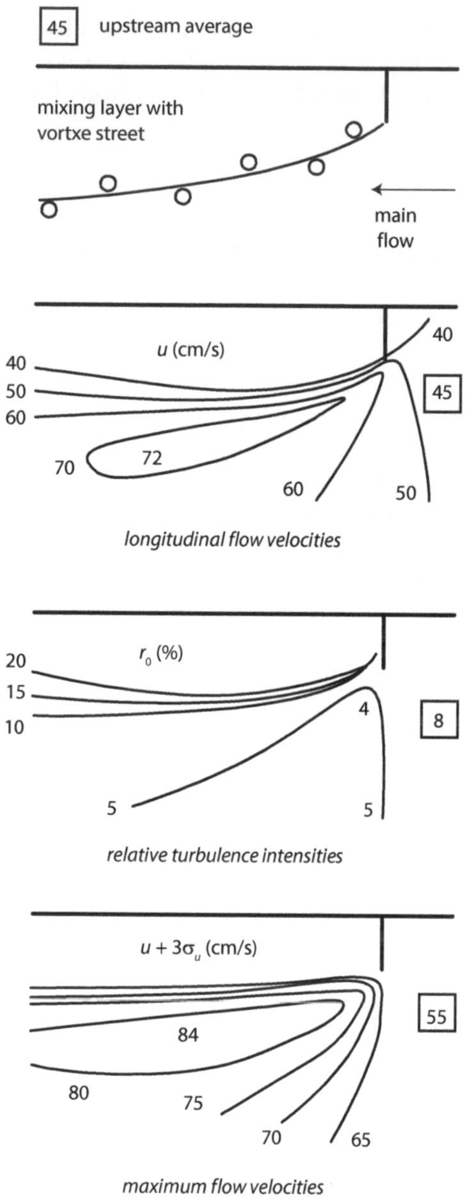

Figure 7.9 Velocities and turbulence intensities for horizontal constrictions (Ariëns, 1993).

models or from experiments in the laboratory. Based on a few flume experiments with horizontal constrictions, Konter and Jorissen (1989) found a relation for the local turbulence coefficient α_l (Figure 7.10). Note that α_l is linked to U_l and α to the mean flow velocity U_0. These values for α_l should be used for a preliminary design and may be valuable for feasibility studies. Scale model investigations are needed for important projects due to the limited predictive power of the equations. If the horizontal

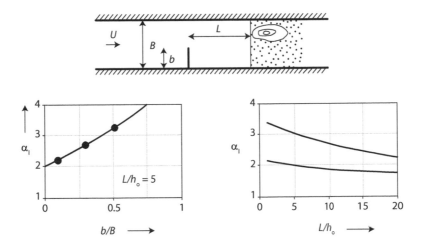

Figure 7.10 Values of α_1 for horizontal constriction.

constriction is marginal compared to the vertical constriction, relations for two-dimensional scour can be applied. It is recommended that the dominant aspect should be selected and that relations for the vertical or horizontal closure should be used.

7.4 Equilibrium scour depth

7.4.1 Introduction

Semi-empirical formulas to estimate the equilibrium scour depth $y_{m,e}$ are given by Inglis (1949), Ahmad (1953), Liu et al. (1961), Veiga da Cunha (1971) and Dietz (1969). Przedwojski et al. (1995) present a review of the application of existing relations to local scour at abutments. Nowadays also a theoretical formula is available by Hoffmans (2012), which is presented in the subsection below as well as a formula that takes the constriction into account too.

Next to this, scour in prototype or in physical small-scale models has been studied. Sieben (RIZA, 2005a) studied the dimensions (depth, volume) of scour holes at groynes in the River Waal. This analysis gives a proper insight into the range of dimensions of the scour holes. Yossef (2002) provides an overview of scouring at impermeable groynes and makes a difference between scour holes for non-overflowing groynes and scour holes for overflowing groynes. At the transition from a non-overflowing groyne to an overflowing groyne, the point where the flow leaves the head of an impermeable groyne changes, and subsequently, the location of the scour hole moves. This is important because most publications emphasise scour holes due to non-overflowing groynes.

Ten Brinke (2003) and Ten Brinke et al. (1999, 2004) describe the erosion by navigation in rivers at low average discharges and sedimentation at high discharges. Apparently, over periods of decades both aspects are in balance. His research was based on neural networks. Berg (2001) describes a research with an adjusted abutment head. Instead of a head with a slope 1V:3H the slope was 1V:6H. It was expected that the scour depth would decrease, but this hypothesis could not be proven. Also in other countries, many studies were carried out and a lot of experiences are available.

For example, much research has been carried out within the framework of the Jamuna Bank Protection Pilot Project FAP 21/22 in the period 1985–1995 (see FAP21/22-RRI, 1993; Klaassen et al., 2012) or for scour at the city of Chandpur, both in Bangladesh (Haskoning, 2002). At Chandpur, the scour hole reached depths up to 70 m.

Three possible erosion or scour types are important for the functioning of an abutment: flow at the root end of the abutments, steepening of the slope of the abutment head, and lowering of the abutment (due to erosion of abutment material, ice, vandalism). These types of erosion or scour result in sedimentation in the main channel.

7.4.2 Calculation methods

In 1938 and 1939, several experiments were carried out at the Central Water Power Research Station, Poona (India) to find detailed information about the scour process near the Harding bridge pier. These experiments formed the basis for the derivation of the Inglis–Poona relation (Inglis, 1949):

$$y_{m,e} + h_0 = 2.32 \left(q^{2/3} / b \right)^{0.78} \tag{7.4}$$

where
    b = width of the abutment (m)
    q = upstream discharge per unit width (m^2/s)

Equation (7.4) is not dimensionally correct and therefore cannot be adopted for general application. Through dimensional analysis, Ahmad (1953) determined the parameters affecting scour depth at abutments and found the following generally applicable relation:

$$y_{m,e} + h_0 = \phi \left(q / (1-m) \right)^{2/3} / g^{1/3} \tag{7.5}$$

where
    ϕ = function of boundary geometry, shape of abutment nose, characteristics of bed material and distribution of velocity in the cross section representing the concentration of flow (–)
    $m = 1 - B_2/B_1$ (–)
    B_1 = width of the river upstream of the abutment (m)
    B_2 = width of the river at the constriction (m)
    g = acceleration of gravity (m/s^2)

In the literature, Equation (7.5) is usually given in the form:

$$y_{m,e} + h_0 = K_A K'_A \left(q / (1-m) \right)^{2/3} \tag{7.6}$$

where
    $K'_A = 2.14 g^{-1/3}$ ($\approx 1.0 \, \text{m}^{-1/3} \text{s}^{2/3}$)
    K_A = correction factor (–) (Table 7.3)

Table 7.3 Correction factors in Ahmad formula

Angle of attack relative to abutment	K_α	Position of structure in bend	K_p
30°	0.80	Straight channel	1.00
45°	0.90	Concave side of bend	1.10
60°	0.95	Convex side of bend	0.85
90°	1.00	Downstream part of concave side	
120°	1.05	Sharp bend	1.40
150°	1.10	Moderate bend	1.10
Structure	K_η	Shape of structure	K_s
20% undrained	1.0	Vertical-wall abutment	1.00
50% porosity (IV:2H)	0.9	Spill-through abutment with 45° side slopes	0.85
50% porosity (IV:3.5H)	0.6		

Equation (7.6) was derived for abutments crossing alluvial rivers in Pakistan and is based on field experience and model studies. The correction factor K_A is a function of the abutment geometry, expressed (within 15% accuracy) as:

$$K_A = 2K_p K_s K_\alpha K_\eta \qquad (7.7)$$

where
   K_p = correction factor for influence of channel bend (–)
   K_s = correction factor for influence of shape of structure (–)
   K_α = correction factor for influence of angle of attack (–)
   K_η = correction factor for influence of porosity (–)

The scour process is governed by effects of asymmetry in the velocity distribution, such as could occur downstream of a bend. The location of abutments within a concave bank alignment is of importance since it can lead to an almost twofold increase in the maximum scour depth as compared to scour caused by an abutment located along a straight channel. The angle of attack of the flow to abutments also influences local scour. Owing to the streamlining effect, the scour depth is reduced for structures angled downstream. Conversely, the scour depth increases if these structures are angled upstream.

The equilibrium scour depth depends significantly on the shape of structure. Therefore, streamlining is effective. The scour depth on the stream side and upstream side of spill-through abutments is about half that of abutments with vertical walls. The scour depth around permeable abutments is less than that around impermeable ones and is strongly dependent on the opening ratio. Detailed studies of scour around permeable abutments were carried out by Orlov, Altunin and Mukhamedov and reported by Przedwojski et al. (1995). River training studies (Halcrow & Partners 1993) in the Brahmaputra river ($y_{m,e} \approx 2$–44 m) provided the K_η-values given in Table 7.3. Some indicative values for the different corrections are given in Table 7.3.

According to Liu et al. (1961), who investigated vertical-wall abutments and spill-through abutments (streamlined structures), the equilibrium scour depth at abutments can be given by:

$$y_{m,e} = K_L h_0 (b/h_0)^{0.4} Fr^{1/3} \tag{7.8}$$

where
Fr = upstream Froude number (–), $Fr = U_0/\sqrt{gh_0}$
K_L = coefficient (streamlined abutments: K_L= 1.1; blunt: K_L =2.15)

Equation (7.8) is based on laboratory experiments but has not been validated for prototype conditions. Note that for blunt abutments the scour depth is about double than for streamlined abutments. Breusers and Raudkivi (1991) re-examined the data of Liu et al. and concluded that for vertical-wall abutments with a negligible thickness, the correction factor is independent of the horizontal constriction and approximates $K_A \approx 2.0$ (Equation 7.7). On the basis of the water continuity equation, Dietz (1969) found the following relation for the equilibrium scour depth for conditions without upstream sediment transport:

$$y_{m,e} + h_0 = \omega \, q/U_c \tag{7.9}$$

where ω = turbulence coefficient (–)
A comparison between the Dietz and Ahmad relations shows a qualitatively different treatment of the influence of the bed material. In the Dietz relation, the bed material is represented in U_c, whereas in Equation (7.6) the bed material is irrelevant. Although the Dietz approach is based on the continuity equation and includes the effects of turbulence at the inflow section, it can be improved by taken into account the influence of turbulence at the outflow that is in the deepest part of the scour hole. The equations of both Dietz and Ahmad can be used for practical applications.

From the results of a scale model investigation with and without bed protection around abutments ($L/h_0 = 0$ and $L/h_0 = 5$), van der Wal (1991) found the following relation for the coefficient ω:

$$\omega = 1.0 + (1.75 + 12m^2)e^{-0.1L/h_0} \quad \text{for} \quad 0.1 < m < 0.5 \tag{7.10}$$

where L is the length of the bed protection parallel to the main flow direction.
The empirical Equation (7.10) can be applied if the abutment has a vertical head or if it is not combined with a sill. The relation of van der Wal was developed for a cohesionless bed material with a uniform gradation. In the experiments where the structure was protected with a bed protection, the flow was hydraulically rough ($k_s/h_0 > 0.025$). If the contraction ratio is larger than 0.5, the extrapolation of Equation (7.10) calculates a further increase in the value of ω, which is not realistic because the discharge through the reduced opening left by the abutment will be greatly reduced. This has been observed in a few tests during the systematic research programme on scouring at Delft Hydraulics (e.g. Delft Hydraulics, 1979).

Ariens (1993) obtained the following relation which is analogous to that of Jorissen and Vrijling (1989):

$$\omega = \alpha_g U_g / U_0 \tag{7.11}$$

where U_g = mean gap velocity (m/s), $U_g = Q/A_g$, α_g = gap coefficient (–), $\alpha_g = 2.4$.

Ariens (1993) found the gap coefficient to be independent of the horizontal constriction and equals about 2.4. The latter value was successfully verified by physical model tests. The inaccuracy of the $y_{m,e}$ value predicted with Equation (7.9), compared to measured values using $\alpha_g = 2.4$, is about 30%, which is considered to be a reasonable result for a preliminary design.

Hoffmans (1995) compared a large number of scour predictors for abutment scour with the experimental data. Extending the Breusers relation with a term for constriction scour proved to give the best results with respect to live-bed scour. The resulting relation is written as

$$y_{m,e} = h_0 \left((1-m)^{-2/3} - 1 \right) + K_B b \tanh(h_0/b) \quad \text{for} \quad U_0 > U_c \tag{7.12}$$

where K_B is a correction factor (Table 7.4). It should be noted that the first term in Equation 7.12 represents scour due to the constriction of the channel by the presence of an abutment or a bridge pier. Applicability of this representation is limited to constrictions which are sufficiently long. For relatively large depths (say $b/h_0 < 1$), the width of the structure will be significant for the scour depth instead of the flow depth. In such cases, the term representing local scour reduces to $y_{m,e} \approx K_B b$, whereas for shallow water conditions $(b/h_0 > 1)$, the second term in Equation (7.12) can be written as $y_{m,e} \approx K_B h$. For circular bridge piers, Breusers found a value of 1.5 for K_B. However, this coefficient is not universal but is strongly dependent on the turbulence

Table 7.4 Correction factor K_B

Type	Code	b/l	α	K_B
Circular pier	CP			1.5
Semi-circular pier	SCP			1.5
	SCE1	<3		1.5
Semi-circular end	SCE2	3–5		2.25
	SCE3	> 5		3.0
Vertical-wall abutment	VWA			3.0
Wing-wall abutment (streamlined)	WWS1	0.2	45°	0.75
	WWS2	0.3	35–45°	1.25
Wing-wall abutment (blunt)	WWB1	0.5–1.5	30°	1.5
	WWB2	1.5–2.5	30°	2.0
Spill-through abutment				
(1.5H:1V)	ST1	0.2		0.75
(1H:1V)	ST2	0.2		1.0
	ST3	0.5–1.5		1.5
Triangular-shaped abutment	TS		45°	1.0

intensity. Depending on the shape of abutment or bridge pier, the value of K_B varied from 0.75 to 3.0. Equation (7.12) depends strongly on the K factor for the geometry, and therefore, it is recommended to apply the equation only in preliminary studies.

Based on a balance of forces, Hoffmans (2012) derived a formula for the scour at abutments for both clear-water scour and live-bed scour. The equilibrium scour for live-bed scour reads:

$$\frac{y_{m,e}}{h_0} = 1.4\chi_e - 1 \quad \text{for} \quad U_0 \geq U_c \tag{7.13}$$

with

$$\chi_e = \frac{1 + 6.3r_0^2}{1 - 6.3r_{0,m}^2} \tag{7.14}$$

where
r_0 = depth-averaged relative turbulence intensity just upstream of the scour hole (–)
$r_{0,m}$ = depth-averaged relative turbulence intensity where the scour depth is at its maximum (–)
χ_e = turbulence parameter (–)

For a first estimate of the equilibrium scour depth, a default value of $r_{0,m} = 0.25$ can be used.

For clear-water scour (usually not relevant for prototype conditions) the formula reads:

$$\frac{y_{m,e}}{h_0} = 1.4\chi_e \left(\frac{U_0}{U_c}\right)^2 - 1 \quad \text{for} \quad U_0 < U_c \tag{7.15}$$

Raudkivi and Melville have been conducting research in Auckland (New Zealand) for decades on the scour process at bridge piers and abutments. Melville and Coleman (2000) provide the most recent update of their design formulas. Their formulas use K factors and provide an upper limit of the scour depth. The approach of Auckland University uses a relation which is based on the Buckingham Pi Theorem. If the abutment is wider than 5 times the flow depth, the upper limit for the maximum scour depth is $y_{m,e} = 4.5h$. If the abutment is 25 times longer than the flow depth, the upper limit for the maximum scour depth is $y_{m,e} = 10h$ which is comparable with an extreme high relative turbulence of 0.37, see also Equation 7.15.

Hoffmans (2012) focuses on a best-guess approach. The ratio between the upper limit of $y_{m,e}$ and the best guess of $y_{m,e}$ is in the extreme case $10h/1.5h = 6.7$ and could probably be ascribed to the magnitude of the turbulence caused by the horseshoe and principal vortices combined with the downflow. Though no turbulence intensities are used for calibration and validation, the modelling here indicates that the maximum scour depth increases if the turbulence intensity increases in the scour hole. The upper limits found by Auckland University could be elucidated by means of a simple turbulence modelling.

Finally, we mention that Galay (1987) presents a summary of design guidelines for spurs in mountainous areas. In China, very often spurs are made of bricks with very steep slopes on a foundation of loose rock. The steep slopes result often in deep scour holes.

7.4.3 Discussion

The designer has to select the method that best suits the scour problem in question. Sometimes it may be necessary to use more than one method and then assess the local scour depth by engineering judgement. The relations discussed in Section 7.4 are generally intended to approximate the scour depth in the equilibrium phase of the scour process. These relations have been developed by using laboratory data and have been verified by some experiments on the prototype scale. Since these relations do not include a safety factor, they are suitable for use in a probabilistic approach to abutment scour. A study of the accuracy of the predictors has been carried out by comparing measured and predicted values of $y_{m,e}$. About 200 experiments concerning scour at abutments were used to verify the aforementioned relations (Hoffmans, 1995). The prediction accuracy of the formula is about 80% for a discrepancy ratio r between the predicted and measured value in the range of 0.75–1.33 for the 200 tests, which corresponds to standard deviation of 0.25μ.

Guidelines of the FHWA (1985) show an extended form of the Ahmad Equations (7.6) and (7.7), adding influences not incorporated in those equations such as factors for the length and width of the bed protection and the porosity of the abutment:

$$\frac{y_{m,e}}{h_0} = f(K_{constriction}, K_{shape\,groyne\,head}, K_{length\,bed\,protection},$$
$$K_{width\,bed\,protection}, K_{flowvelocity}, K_{orientation\,groyne}) \qquad (7.16)$$

The equation resembles a relation between the maximum scour depth at spill-through abutments developed by Haskoning (2006), but the structure is also common for scour formulas; see, for example, the formulas of Liu et al (1961) and Hoffmans (2012). The formula is based on an analysis of many physical model tests within Dutch fundamental research (WL|Delft Hydraulics, 1972a; WL|Delft Hydraulics, 1972b; WL|Delft Hydraulics, 1991) and research in the framework of river training works in Bangladesh (FAP21/22-RRI, 1993). However, Equation (7.16) is difficult to use because for only some of the K parameters values are available.

7.5 Combined scour

7.5.1 Introduction

The scour of the bed near bridge abutments and bridge piers in rivers can be considered as a composite of bed scour by different processes like overall degradation, constriction scour, bend scour and local scour (Section 4.2). Overall degradation is caused by the natural changes in the stream and constriction scour is caused by the narrowing of the waterway at the site. The spiral flow in river bends leads to bend scour in the outer bend. Local scour around abutments results from flow disturbances introduced by the presence of the structure. As a first estimate, the scour which is caused by each separate process may be added linearly to obtain the resulting scour.

7.5.2 Combined local scour and constriction or bend scour

Local scour generally coincides with general scour such as confluence scour and bend scour. Calculation methods for constriction and bend scour are presented. More general prediction relations for general scour are given in Section 4.2.

Hoffmans and Buschman (2018) present a formula for the equilibrium depth of combined scour due to constriction and the presence of the abutment:

$$\frac{y_{m,e}}{h_0} = \left\{ \left(1 - \frac{A_a}{A_r}\right)^{-\frac{2}{3}} - 1 \right\} + \left\{ \alpha_{ga} \frac{1 + 6.3 r_0^2}{1 - 6.3 r_{0,m}^2} - 1 \right\} \tag{7.17}$$

where
   A_a = cross-section area of the abutment (m^2)
   A_r = cross-section area of the river (m^2)

For non-submerged abutments $\alpha_{ga} = 1.4$ holds; otherwise for submerged abutments $\alpha_{ga} = 1.0$. In all the tests, r_0 was assumed to be equal to $r_0 = 0.1$ and $r_{0,m}$ is calculated by using Equation (7.18) with $c_3 = 0.33$ (Hoffmans & Booij, 1993a, b):

$$r_{0,m} = c_3 \sqrt{\frac{A_b}{A_f} \left(\frac{L}{6.67 h_0} + 1\right)^{-1.08}} \tag{7.18}$$

where
   A_b = two-dimensional blocking area of the abutment (m^2)
   A_f = cross-section area upstream of the abutment (m^2)
   L = the length of the bed protection in streamwise direction (m)

About 50 flume experiments were applied for validating Equation (7.17). About 90% of the tests predicted a scour depth that varies from 0.5 to 2.0 times the measured scour depth, implying a standard deviation of 0.25 m. In Section 7.7, an example is treated.

The combined scour depth of local scour at an abutment and bend scour in a channel meander can be determined by superposition of the individual scour depths (Figure 7.11). Mesbahi (1992) found that the total scour depth behind an abutment in a channel bend equals the sum of the bend scour A and the local scour B. The local scour is computed by Equation (7.6), based on the average flow depth h_0, and not on the local depth $h_0 + A$. In general, the earlier presented relation of Ahmad Equation (7.6) includes bend scour.

7.6 Failure mechanism and measures to prevent local scour

7.6.1 Introduction

A scour hole just downstream of the head of abutments can endanger the stability of the structure. In the Netherlands, much experience has been obtained from the use of additional protection of the scour hole to prevent excessively deep scouring and

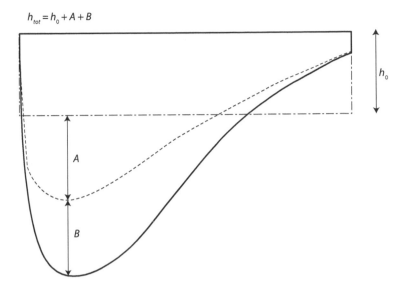

$h_{tot} = h_0 + A + B$

h_0

A

B

h_0 = average depth in straight channel without abutment

A = bend scour (without abutment)

B = local scour due to abutment in straight channel

Figure 7.11 **Combined local scour and bend scour.**

undermining of groynes along the rivers Rhine, Meuse and IJssel. This section considers the scour depth near an isolated abutment attacked by a more or less uniform and steady approach flow. It does not consider the possible interaction between the scour holes of a series of groynes in a row along a bank.

Countermeasures can be divided in the following types (for more details, see Section 2.4):

1. Bank hardening countermeasures
2. Embankment stabilising countermeasures
3. Flow altering countermeasures.

7.6.2 Scour slopes

The initial slope of a scour hole near an abutment cannot be estimated with the aid of the relations currently in use. Data of the initial slopes measured in scale model tests are available, for example, from systematic research tests (Delft Hydraulics, 1972, 1979; van der Wal, 1991). These experimental results could be very useful to researchers attempting to derive relations for the estimation of the initial slope of a scour hole. In general, the stability of non-submerged abutments is endangered by scour holes due to the strong vortices just downstream of the separation point at the head of the abutment. Too steep a scour hole may induce part of the head to slide into the scour hole or may cause undermining of this head. The soil mechanical aspects of a failure caused by a scour hole that is too steep, and the consequent damage to the structure, are treated by Silvis (1988) (Chapter 3). The crest of a submerged abutment may be damaged by a supercritical flow and a hydraulic jump. In that situation, special attention must be paid to the design of the crest protection.

7.6.3 Outflanking

A frequent mode of abutment failure is outflanking. Outflanking is defined as a breach in the connection between the abutment and the bank line creating a connection between two abutment fields if the water level is high enough. A consequence can be erosion of the causeway on top of the embankment while the abutment structure remains intact. The progressive development of an eddy in this recess draws the interface back from the head of the abutment into the stagnation region, so that the downflow strength increases and the scour depth increases. The result may be a breakthrough behind the abutment head. In exceptional cases, the scour hole due to the vortex street downstream (see Figure 7.12) and the deceleration of the flow can endanger the stability of the downstream wing-wall. The formation of such a scour hole can be prevented by an extended bed protection, or by an increase of angle α_2. Alternatively, the stability of the wing-wall can be guaranteed by increasing its foundation depth.

In principle, outflanking has to be prevented because it has many undesired effects: disturbed flow pattern and sedimentation. Therefore, quick action to repair a breach is required. The damage caused by the undermining of wing-walls can be repaired by taking the measures mentioned in Chapter 8 (Bridge Piers). Figures 7.12–7.14 are examples of outflanking with a velocity of about 10 m/year. More information can be found in RIZA (2005b).

Figure 7.12 **Observed outflanking.**

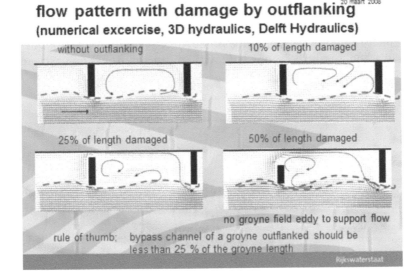

Figure 7.13 **Flow pattern during outflanking.**

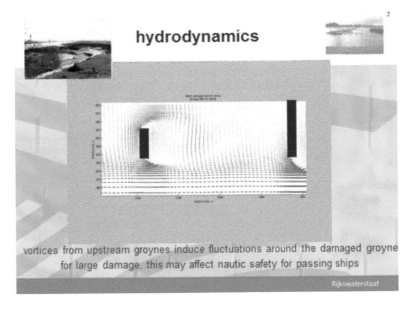

Figure 7.14 **Vortices around damaged abutment.**

7.6.4 Riprap protection

The scour hole due to the general flow acceleration and the vortex street downstream
of the separation point endangers the stability of the side wall of an abutment without an adjacent bed protection. When the side wall consists of vertical sheet piles,

this scour hole may have a considerable depth. This depth will be reduced by sloping the side wall, while a bed protection can prevent the formation of such a scour hole (Figure 7.15).

The foundation of a wing-wall abutment can be strengthened by a bed protection to prevent the formation of a scour hole in the direct vicinity of the abutment. However, in that situation a reduced scour hole will be generated just downstream of the bed

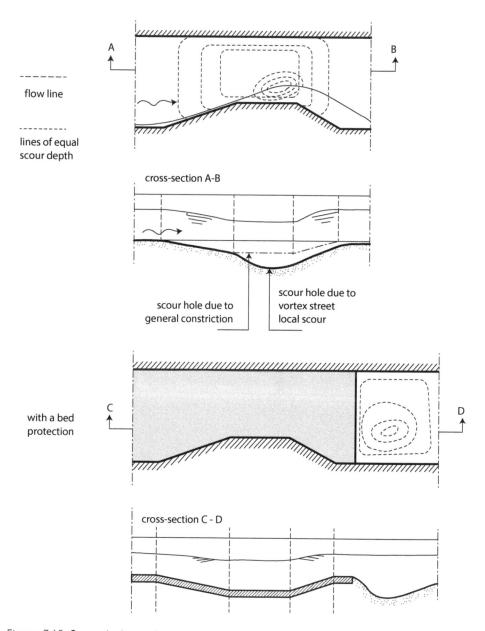

Figure 7.15 Scour hole at abutment.

protection. When expecting a strong flow attack on the head of an abutment, the head is usually strengthened by making it bigger. Other possible measures are guide vanes or sills (Chapter 6).

7.7 Examples

7.7.1 Introduction

Two examples of scour are addressed. In Section 7.7.2, the scour due to the lowering of existing abutments is computed. In particular, the effect of the relative turbulence intensity is shown. The results are compared with measured scour depths. Section 7.7.3 discusses the effect on the scour depth of a permeable abutment compared to the scour depth resulting from an impermeable abutment.

7.7.2 Scour due to lowering of existing abutments

Within the framework of 'Space for the Rhine Branches' (a program commissioned by the Dutch Ministry of Transport, Public Works and Water Management to lower design flood levels and increase spatial quality), various measures have been developed to decrease the flow levels at peak discharges. One of those measures is lowering the existing abutments. We briefly discuss the consequences for scour.

The water depth in the Waal varies between approximately 4.0 m during low flow and 7.0 m during floods. The depth-averaged relative turbulence intensity $r_0 = 0.10$ in the Waal. Assuming $r_{0,m} = 0.25$, the turbulence parameter $\chi_e = 1.75$ and $y_{m,e} = (1.4 \cdot 1.75 - 1)$ $h = 1.45h$. This maximum equilibrium depth of the scour hole can be compared to the measured scour holes behind the abutments in the Waal (Vrolijk et al., 2005). Table 7.5 shows the results of these field measurements along with the y_m/h_0 relation and the $r_{0,m}$ that would result in an equilibrium scour hole depth and the measured scour hole

Table 7.5 Measured scour depths

Date	Q (m^3/s)	h_0 (m)		y_m (m)	y_m/h_0 (-)	$r_{0,m}$ (-)
25-11-2002	3300	6.6	high water	2.4	0.36	-
				2.1	0.32	-
				2.2	0.33	-
				2.6	0.39	-
18-03-2003	1710	4.3	low water	2.8	0.65	0.125
				2.6	0.60	0.107
				3.8	0.88	0.183
				3.3	0.77	0.158
17-02-2005	3650	6.7	high water	3.2	0.48	-
				2.8	0.42	-
				3.0	0.45	-
				3.3	0.49	-
26-05-2005	1570	4.1	low water	3.4	0.83	0.172
				2.7	0.66	0.128
				3.5	0.85	0.177
				3.4	0.83	0.172

depth. For high water conditions, it is apparent that the equilibrium scour hole depth has not been reached as there is no non-negative $r_{0,m}$ that would return the measured scour hole depth.

This observation is consistent with the evaluation of the appropriate time scales of scouring and water movement. The characteristic time (t_1) at which $y_m = h_0$ reads (e.g. van der Wal, 1991)

$$t_1 = \frac{K_{330}h^2\Delta^{1.7}}{(\alpha U_0 - U_c)^{4.3}} \quad \text{(in hours)}$$

where K_{330} [$=330$ hours m$^{2.3}$/s$^{4.3}$] is a coefficient, α ($=1.5 + 5r_0$) is a turbulence coefficient (Jorissen & Vrijling, 1989) and Δ is the relative density. With $h_0 = 6.5$ m, $r_0 = 0.1$, $U_0 = 1.2$ m/s and $U_c = 0.7$ m/s, t_1 is about 3300 hours, i.e. 20 weeks. A period of high water lasts about two weeks, which means that the equilibrium value will certainly not be achieved during high water conditions, as is apparent in Table 7.5.

For low water, the equilibrium depth would be expected to be reached as the water level is always higher or equal to the low water level. This means that from the measurements, it can be concluded that for abutments in the Waal turbulence levels at the maximum scour depth $r_{0,m}$ between 0.15 and 0.20 are appropriate and that the suggested 0.25 is a safe assumption for design purposes.

Figure 7.16 shows the dependence of the maximum equilibrium scour hole depth $y_{m,e}$ for a water depth $h_0 = 4.1$ m and $r_0 = 0.1$. The green dots indicate the measurements for low water on 26 May 2005 in the table. The graph shows that the higher the turbulence intensity $r_{0,m}$, the more sensitive the equilibrium scour depth $y_{m,e}$ is. Additionally, to the relative turbulence intensity at the scour depth maximum $r_{0,m}$, the relative turbulence intensity just upstream of the scour hole r_0 might differ from the assumed 0.10.

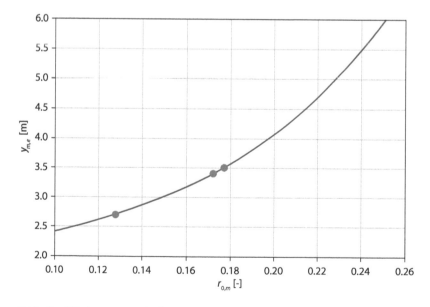

Figure 7.16 Equilibrium scour hole at abutment.

7.7.3 Influence of the permeability of an abutment on the scour

Hoffmans and Buschman (2018) present a formula for the equilibrium depth of combined scour due to constriction and the presence of the abutment:

$$\frac{y_{m,e}}{h_0} = \left\{ \left(1 - \frac{A_a}{A_r}\right)^{-\frac{2}{3}} - 1 \right\} + \left\{ \alpha_{ga} \frac{1 + 6.3 r_0^2}{1 - 6.3 r_{0,m}^2} - 1 \right\} \tag{7.17}$$

For non-submerged abutments $\alpha_{ga} = 1.4$ holds. It is assumed that $r_0 = 0.1$ and $r_{0,m}$ is calculated by using Equation (7.19) with $c_3 = 0.33$:

$$r_{0,m} = c_3 \sqrt{\frac{A_b}{A_f} \left(\frac{L}{6.67h} + 1\right)^{-1.08}} \tag{7.18}$$

We assume a situation without a bed protection, thus $L = 0\,\text{m}$. The ratio A_b/A_f represents the permeability of the abutment (see Figure 7.17). If $A_b/A_f = 1$, then the abutment is impermeable and with Equation (7.18) this results in $r_{0,m} = 0.333$. Suppose we have an abutment with a permeability of 50% or $A_b/A_f = 0.5$, this results in $r_{0,m} = 0.233$.

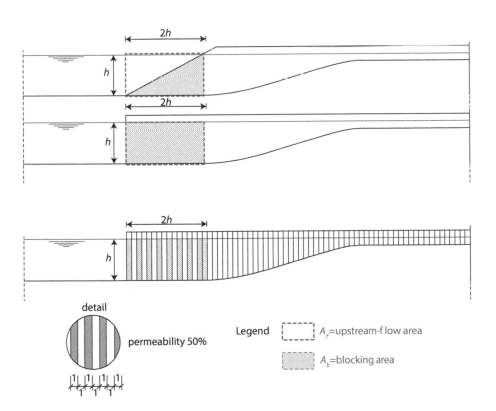

Figure 7.17 Definitions of A_f and A_b for abutments of different shapes and permeability.

Suppose we have a water depth of 5 m, a river width of 100 m and a abutment width of 25 m, this results in an abutment area $A_a = 125\,\mathrm{m}^2$ and a river area $A_r = 500\,\mathrm{m}^2$.
Substituting these values in Equation (7.17) results in:

- impermeable abutment: $y_{m,e} = 25.75\,\mathrm{m}$
- 50% permeable abutment: $y_{m,e} = 12.35\,\mathrm{m}$

It can be concluded that a permeable, non-submerged abutment reduces scour depths considerably compared to a non-permeable abutment.

Chapter 8

Bridges

8.1 Introduction

Bridge piers produce a horizontal constriction of the flow, comparable to that caused by abutments. The local scour around bridge piers depends strongly on the geometry of the pier. The characteristic flow pattern at structures in relatively deep water as well as at submerged bridges is described in Section 8.2. As in the previous chapters, some aspects of the time-dependent development phase of the scour process is treated in detail, with the objective of extending the applicability of the Breusers method (Section 8.3). Local scour at a bridge pier usually evolves rapidly. Therefore, most investigators are interested in the maximum scour depth in the equilibrium phase (Section 8.4). Distinctive parameters of bridge piers that influence the scour process are discussed in Section 8.5. The schematised shape of local scour is determined by the maximum scour depth and the side slopes of the scour hole. Section 8.6 gives some information about making a first estimate of the shape of a schematised scour hole. Various failure mechanisms of bridge piers are illustrated from experience with practical examples, and some methods to prevent local scour are also mentioned (Section 8.7). Section 8.8 presents an example regarding the differences between slender and wide piers. Many publications are available that give specific information about the local scour near bridge piers. It is beyond the scope of this manual to give a complete review of all these publications. Nevertheless, one particular issue regarding bridge scour is addressed, viz pressure scour where the bridge deck influences the scour depth.

8.2 Characteristic flow pattern

8.2.1 Introduction

The flow pattern near a pier is rather complicated (Figures 8.1 and 8.2). This complex flow pattern in the scour hole has been described in detail by several authors, for example Herbich (1984), Dargahi (1987), Breusers and Raudkivi (1991) and NCHRP (2011). The main flow features causing scour are the wake vortices and the horseshoe vortices combined with the average downflow (see Figure 8.1). The axis of the horseshoe vortex is horizontal and wraps around the base of the pier in the shape of a horseshoe. The wake-vortex system has vertical axes, which are commonly seen as eddies. The geometry of the pier significantly influences scour depth because it reflects the load due to the horseshoe vortex at the base of the pier. Here, attention is paid to local scour around

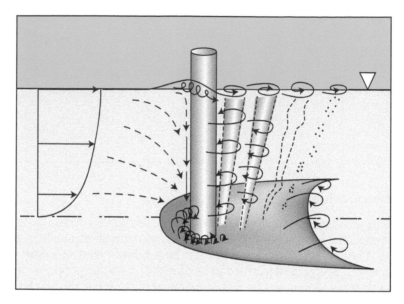

Figure 8.1 The main flow features forming the flow field at a slender pier of circular cylindrical form (*b/h* < 1) (NCHRP, 2011).

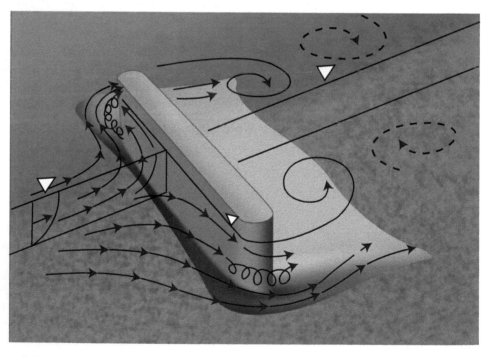

Figure 8.2 Main features of the flow field at a wide pier (*b/h* > 1) (NCHRP, 2011).

slender bridge piers, i.e. the ratio between the width (b) of the pier and flow depth (h_0) is smaller than 1. Calculation methods for geometries where $b/h_0 > 1$, for example scour at abutments, are discussed in Chapter 7.

8.2.2 Submerged bridges

Before discussing the calculation methods to determine scour depth at bridge piers some attention has to be paid to pressure scour. This type of scour may occur during high discharges at bridges with relatively low superstructures or bridge decks. The water flows against the superstructure (which also allows floating debris to concentrate upstream). Because the water below the bridge accelerates, it exerts a higher load on the bed and subsequently more scour.

This type of scour rarely occurs in the Netherlands, however there are two examples. First, a bridge over a ditch in the Meuse floodplain at Roosteren, against which floating debris piled up during the flood in 1995. The bridge collapsed, and the Meuse started to cut off its bend along this itinerary. Second, the condition of submerged flow has been taken into account for design flood conditions (Maatgevend HoogWater (MHW)) on the Nederrijn at Meinerswijk. Nevertheless, bridges in the Netherlands are usually designed with the assumption of open-channel flow condition, but the flow regime can switch to pressure flow when the downstream edge of a bridge deck is partially or totally submerged during a large flood. Figure 8.3 shows at the left hand, a bridge undergoing partially submerged flow in Salt Creek, United States, in June 2008. At the right hand of Figure 8.3, a totally submerged flow is shown in Cedar River, United States, in June 2008, which interrupted traffic on highway I-80.

Unlike open-channel flows, these pressure flows create a severe scour potential. Wakes lead to significant turbulence levels, and the highly turbulent pressurised flow causes scouring of the channel bed. Three types of pressure scour can be distinguished (Case I, Case II and Case III; see Figure 8.4). Note that there are no bridge piers situated in the channel. They would increase the scour. In Section 8.4.3, a computation method for each case will be presented.

Figure 8.3 **Partially submerged bridge (left) and a fully submerged bridge (right)** (FHWA, 2009).

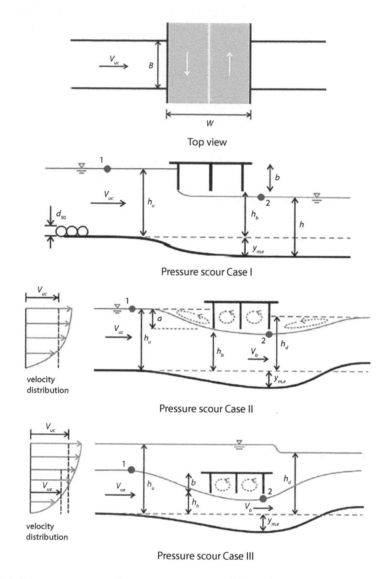

Figure 8.4 **Pressure scour: relevant parameters and flow characteristics (FHWA, 2009).**

8.3 Time scale

The process of local scour around bridge piers can be divided into several phases: initial phase, development phase, stabilisation phase and equilibrium phase (Section 3.4.4). In general, local scour around bridge piers evolves rapidly. For engineering purposes, a time-dependent description of the maximum scour depth is not always relevant. However, in rivers with relatively large flow depths, the time factor can be important, for example, during the construction phase. The Breusers method was originally developed for sills with a bed protection (Chapter 6). In the 1970s, successful attempts were made to apply the same approach to the scour process around bridge piers (e.g. Breusers, 1971; Akkerman, 1976; Konter, 1976, 1982; Nakagawa & Suzuki, 1976). In this section, the results of these efforts are summarised.

For slender bridge piers ($b/h_0 < 1$), the maximum scour depth can be described by the following relation, which is valid for all phases of the scour process, provided $y_{m,e} > b$:

$$\frac{y_m}{y_{m,e}} = 1 - e^{-\frac{b}{y_{m,e}}\left(\frac{t}{t_1}\right)^{\gamma}} \tag{8.1}$$

where
  b = width of the pier (m)
  t = time (s)
  t_1 = characteristic time at which $y_m = b$ (s)
  y_m = maximum scour depth at t (m)
  $y_{m,e}$ = maximum scour depth in the equilibrium phase (m)
  γ = coefficient (-), $\gamma = 0.2$–0.4

In the development phase, that is for $t < t_1$, Equation (8.1) reduces to (Figure 8.5):

$$\frac{y_m}{b} = \left(\frac{t}{t_1}\right)^{\gamma} \tag{8.2}$$

Equation (8.2) has not been investigated thoroughly in the Netherlands for scour at bridge piers, however experiments were carried out in Japan. According to Nakagawa and Suzuki (1976), the value of the coefficient $\gamma = 0.22$–0.23 and the characteristic time t_1 could be written as (Figure 8.6):

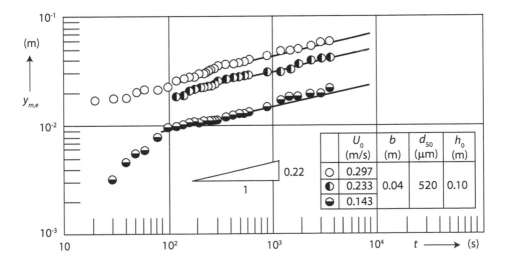

Figure 8.5 Time variation of maximum scour depth.

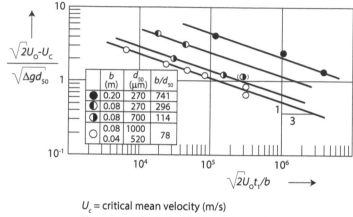

U_c = critical mean velocity (m/s)

U_o = mean velocity (m/s), $U_o = Q/A$, Q is discharge (m³/s), A is cross section (m²)

Δ = relative density (–)

Figure 8.6 Characteristic time t_1

$$t_1 = 29.2 \frac{b}{\sqrt{2}U_0} \left(\frac{\sqrt{\Delta g d_{50}}}{\sqrt{2}U_0 - U_c} \right)^3 \left(\frac{b}{d_{50}} \right)^{1.9}$$

(8.3)

where

d_{50} = mean particle diameter (m)
g = acceleration of gravity (m/s²)
U_c = critical mean velocity (m/s)
U_0 = mean velocity (m/s)

Equation (8.3) has been derived in laboratory tests, but it has not been validated in prototype conditions, and thus it is not generally applicable. We emphasise the resemblance between Equation (8.3) and Equation (6.9), including the values for the coefficients in Table 6.2. Obviously, h_0 is replaced by b if b/h is greater than 1.

Nakagawa and Suzuki (1976) studied the characteristics of local scour around bridge piers with rectangular cross-sections in a tidal current ($0.04 < b < 0.20$ m). About 30 flume experiments were carried out in which the flow direction and the wave attack varied. Several sands were used ($270 < d_{50} < 1000$ μm). The flow velocities varied from 0.13 to 0.30 m/s, and the flow depth was 0.10–0.20 m. In addition, a steel cylinder with $b = 9$ m diameter was tested in a prototype experiment. The tests showed that the scour hole which develops from a corner of a rectangular pier does not change in form. The test results in which $b/h_0 > 1$ (e.g. Konter, 1976, 1982) also suggest that the Breusers approach is applicable, viz. Equation (8.2) with a value for γ of 0.4 and for the time t_1 Equation (8.3). In the case of tidal flow, the maximum scour depth decreases during a reverse-flow period. The decrease is marginal compared to the equilibrium scour depth.

8.4 Equilibrium scour depth

8.4.1 Introduction

Several investigators examined the scour process around bridge piers. Scour relations were derived purely from curve fitting (e.g. Hopkins et al., 1980; Breusers et al., 1977). Hoffmans (1995) conducted a desk study in which more than 60 scour relations were tested using approximately 1000 flume experiments. This section, however, only discusses those expressions of the equilibrium scour depth that pertain to the predictors with the highest accuracy. Recently, Hoffmans (2012) developed a method based on a balance of forces.

8.4.2 Calculation methods

Laursen and Toch (1956) based a scour relation on a conservative curve, drawn free-hand, to model study data for the sediment-transporting case. They also presented multiplication factors for the shape or angle of attack. They investigated the influence of pier shape, angle of attack, flow depth, velocity and sediment size. As might be expected in view of the large projected width of the pier, flow depth has some influence on the process. Based on the data of Laursen and Toch (1956), Breusers et al. (1977) found a relation for circular bridge piers, which is expressed as:

$$y_{m,e} = 1.35 K_i b^{0.7} h_0^{0.3} \tag{8.4}$$

where K_i = correction factor; see Section 7.5 (for circular piers: $K_i = 1.0$).

For live-bed scour, the equilibrium scour depth fluctuates with time. Therefore, Breusers et al. (1977) recommended taking $y_{m,e} + 0.5\,m$ dune height for design purposes. Based on model studies of drilling spuds in waves and current in the early 1960s, Breusers deduced the following relation to predict the scour depth:

$$y_{m,e} = 1.4b \tag{8.5}$$

This best-guess predictor is applicable to prototype conditions and live-bed scour.

For live-bed scour, Tsujimoto (1988) arrived at a similar relation ($y_{m,e} = 1.42b$). They described the scour process around a circular pier using a model based on reasonable understanding of its essential mechanism. Breusers et al. (1977) concluded from the experimental data that for live-bed scour, thus $U_0 > U_c$, the average scour depth around circular piers may be described by the following relation:

$$y_{m,e} = 1.5 K_i b \tanh\left(h_0/b\right) \tag{8.6}$$

Equation (8.6) gives a good description for the full range of b/h_0 (Chapter 7) and is applicable for live-bed scour in prototype conditions, provided that the correct values for the K factors are used. For slender bridge piers and $K_i = 1$, the equation reduces to $y_{m,e} = 1.5b$ and for wide bridge piers to $y_{m,e} = 1.5h_0$. Note that Equation (8.6) can also be applied to abutments. Then, a simplified version reads also $y_{m,e} = 1.5h_0$ (1.5 is the average value; the range is 0.75–3.0).

For clear-water scour conditions where $0.5 < U_0/U_c < 1$, Breusers et al. (1977) proposed an upper-limit formula:

$$y_{m,e} = 2K_i b (2U_0/U_c - 1) \tanh(h_0/b) \tag{8.7}$$

Equation (8.7) is generally applicable with correct K factors (see tables and graphs in Section 8.5). No scouring occurs if the approach flow velocity is smaller than half the critical flow velocity.

For live-bed conditions, the equation becomes:

$$y_{m,e} = 2K_i b \tanh(h_0/b) \quad \text{for} \quad U_0/U_c > 1 \tag{8.8}$$

Here, the constant 1.5 has been replaced by 2 for a more conservative design, and thus a safety factor of 1.33 ($= 2.0/1.5$). It is recommended to relate the value of this constant to the failure of probability (see also Section 2.3). Johnson (1992) introduced a safety coefficient related to the failure probability (Table 2.2).

One of the more commonly used models in the United States is the HEC-18 formula, also called the Colorado State University (CSU) relation (e.g. Johnson, 1992). This relation was developed following both a dimensional analysis of the parameters affecting pier scour and an analysis of laboratory data. The result was a regression relation consisting of dimensionless ratios. For circular bridge piers, the Colorado State University relation is as follows:

$$y_{m,e} = 2.0 K_i h_0 Fr^{0.43} \left(\frac{b}{h_0}\right)^{0.65} \tag{8.9}$$

where Fr ($= U_0/(gh_0)^{1/2}$) is the upstream Froude number, and $K_i = K_1 K_2 K_3$ (correction factors for pier nose shape, angle of attack of flow and bed condition, respectively; values are presented in Section 8.5). The coefficient determined by regression was modified to 2.0 in order to develop an envelope curve. Therefore, Equation (8.9) incorporates a factor of safety by overpredicting the pier scour depth.

Furthermore, the Liu et al. (1961) formula should be mentioned which originally was derived for abutments (see Chapter 7):

$$y_{m,e} = K_L h_0 (b/h_0)^{0.4} Fr^{1/3} \tag{7.8}$$

According to Liu values for K_L are:
   $K_L = 1.1$ for streamlined abutments, and
   $K_L = 2.15$ for blunt abutments.

Many bridges in the US showed signs of weakening due to scour around the pier foundations. To design safe bridge piers, such that the foundations will not be undermined by scouring, Johnson (1992) developed a probabilistic approach to pier scour engineering. His best-fit model is similar to the CSU relation:

$$y_{m,e} = 2.0 K_i h_0 Fr^{0.21} \sigma_g^{-0.24} (b/h_0)^{0.98} \tag{8.10}$$

where σ_g represents the sediment gradation (i.e. d_{84}/d_{50}). Hydraulic variables, such as discharge, flow depth and flow velocity, are generally stochastic in nature.

Because these parameters determine the scour process, the scour depth is a stochastic variable too. Based on a Monte Carlo simulation, Johnson found a relation between safety factors and the probability of bridge pier scour (Table 2.2). For example, if all bridges in a certain region are required to be designed for a risk level of 10^{-4}, Equation (8.10) can be multiplied by a safety factor of 1.75.

Hoffmans (2012) describes the scour process at bridge piers. He derives a formulation based on a balance of forces for both clear-water scour and live-bed scour; see Section 3.4.3 for the general formula for sills and bridge piers and Table 3.1 for constants valid for bridge piers. For clear-water scour the formula reads for slender bridge piers:

$$\frac{y_{m,e}}{b} = 1.6\chi_e\left(\frac{U_0}{U_c}\right)^2 - 1.3 \quad \text{for} \quad U_0 < U_c \tag{8.11}$$

For live-bed scour, the formula is:

$$\frac{y_{m,e}}{b} = 1.6\chi_e - 1.3 \quad \text{for} \quad U_0 \geq U_c \tag{8.12}$$

with

$$\chi_e = \frac{1 + 6.3r_0^2}{1 - 6.3r_{0,m}^2} \tag{8.13}$$

where
r_0 = depth-averaged relative turbulence intensity upstream of the scour hole (-)
$r_{0,m}$ = depth-averaged relative turbulence intensity at the location with maximum scour depth (-)
χ_e = turbulence parameter (varies between 1.7 to 2.3) (see Figure 8.7, Hoffmans, 2012)

Coefficient values for wide bridge piers are not available (see Table 3.1). However, an estimate can be obtained using Equation (8.14) which is the same as Equation (7.13) for abutments:

$$\frac{y_{m,e}}{h_0} = 1.4\chi_e - 1 \quad \text{for} \quad U_0 \geq U_c \tag{8.14}$$

Equations (8.11) to (8.14) are based on a balance of forces and theoretically correct. However, they have not been validated for prototype conditions, and hence they are not generally applicable. For a first estimate of the equilibrium scour depth, the following values of the relative turbulence intensities can be used: $r_0 = 0.1$ and $r_{0,m} = 0.25$ (thus, $\chi_e = 1.75$). This results for live-bed scour in $y_{m,e} = 1.5b$. If more information on the turbulence field is available (for example, as a result of turbulence models or a physical model study), the turbulence parameter (χ_e) can be determined more accurately (see Section 3.4.3).

Raudkivi (see Breusers & Raudkivi, 1991) and Melville (see Melville & Coleman, 2000) conducted research in Auckland (New Zeeland) on the scour process at bridge piers and abutments. Melville and Coleman (2000) present the most recent update of their design formulas with K factors providing an upper limit of the scour depth (see Section 7.4).

The approach of Auckland University uses a relation which is based on the Buckingham Pi Theorem. For $b/h < 0.7$ (b/h is the ratio between the width of the pier and flow depth) Melville and Coleman (2000) found for $y_{m,e}$:

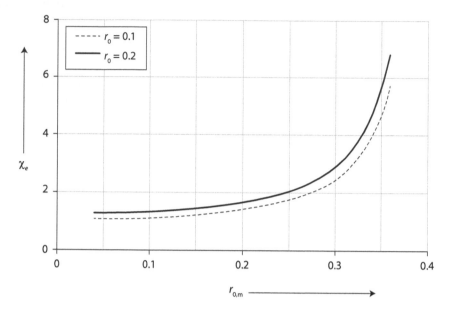

Figure 8.7 Turbulence parameter as a function of the relative turbulence intensity.

$$\frac{y_{m,e}}{b} = 2.4K_{M,p} \qquad (8.15)$$

where $K_{M,p}$ represents an overall correction for flow intensity, pier shape and sediment size, and $y_{m,e}$ is the maximum scour depth. The correction factors are obtained from laboratory experiments and are quantified by using graphs and tables.

According to Auckland University $y_{m,e}$ could reach up values of 2.4 times the width of the pier for clear-water scour, which is about 1.5 times the mean value. The magnitude of $y_{m,e}/b$ is influenced by the turbulence both upstream of the scour hole and in the scour hole itself. If only bed turbulence is considered and the bed becomes rougher, for example r_0 varies from 0.08 to 0.12, the maximum scour depth lies in the range of $1.4b$ to $1.6b$ by using Equation (8.12). If r_0 increases from 0.1 to 0.2, as a result of turbulence caused by vortices with horizontal and vertical axes, $y_{m,e}/b$ increases from 1.5 to 2.0 (see Figure 8.8).

The depth-averaged relative turbulence intensity in the scour hole represents the turbulent energy generated by the horseshoe vortices. If $r_{0,m}$ is larger than $r_{0,m} = 0.25$, e.g. $r_{0,m}$ varies from 0.25 to 0.30, then it follows from Equation (8.12) that the maximum scour depth is $1.5 < y_{m,e}/b < 2.6$. In this way, the upper limit ($y_{m,e}/b = 2.4$ for circular piers) given by Melville and Coleman can be explained (see also Hoffmans, 2012).

Pier scour in cohesive materials generally progresses more slowly and is more dependent on soil properties than pier scour in non-cohesive sediments. Briaud et al. (2001) present a pier scour equation for cohesive material that incorporates the critical velocity (U_c) for initiation of erosion. Based on flume experiments they found for the equilibrium scour depth:

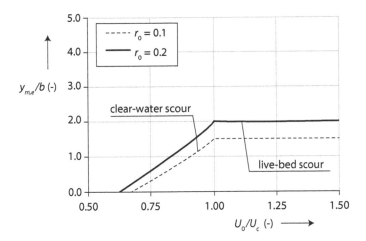

Figure 8.8 Turbulence parameter as a function of the relative turbulence intensity.

$$y_{m,e} = 2.2 K_i b^{0.65} \left(\frac{2.6 U_0 - U_c}{\sqrt{g}} \right)^{0.7} \tag{8.16}$$

where K_i represents the influence of various parameters.

8.4.3 Pressure scour

Pressure scour has been investigated in US research (FHWA, 2009). Figure 8.9 shows the experimental results of the investigations including an equation for a bridge without piers in the channel. The parameters have been explained in Figure 8.4. The formula for the equilibrium scour depth reads:

$$y_{m,e}\left(Fr_s, h_b, a\right) = \left(h_b + a\right)\sqrt{\frac{1 + 1.3680/Fr_s^{2.409}}{1 + 1.8652/Fr_s^2}} - h_b \tag{8.17}$$

with

$$Fr_s\left(V_{uc}, h_u, h_b\right) = \frac{V_{uc}}{\sqrt{g\left(h_u - h_b\right)}} \tag{8.18}$$

$$V_{uc} = 7.69 \cdot \sqrt{\Psi_c \left(s - 1\right) g d_{50}} \cdot \left(\frac{h_u}{d_{50}}\right)^{1/6} \tag{8.19}$$

Figure 8.9 shows Equation (8.17) in another form:

$$\frac{h_b + y_m}{h_b + a} = \sqrt{\frac{1 + 1.3680/Fr_s^{2.409}}{1 + 1.8652/Fr_s^2}} \tag{8.20}$$

In case 3 when water is flowing over the top of the bridge deck (see Figure 8.4), an effective velocity V_{ue} should be applied:

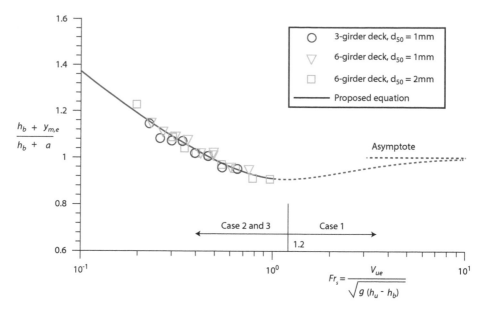

Figure 8.9 Methods to compute pressure scour (see Figure 8.4 for the cases).

$$V_{ue} = V_{uc}\left(\frac{h_b + a}{h_u}\right)^{0.85}$$

(8.21)

where
    V_{uc} = critical velocity (m/s)
    V_{ue} = effective critical velocity in case of case 3 (m/s)
    Ψ_c = critical Shields parameter (-)
    s = the specific density and equals s = 2.65.
    d_{50} = median diameter (m)
    h_u = upstream water depth (m)
    h_b = water depth under bridge (m)
    a = deck block height (m)
    y_m = scour depth (m)
    $y_{m,e}$ = equilibrium scour depth (-)
    Z = scour number (-)

8.4.4 Discussion

Some of the relations mentioned above are for clear-water scour, some are for live-bed scour and some are intended to serve for both. These relations are usually intended to estimate the ultimate scour depth that can be expected under the conditions of a specific design flow velocity or depth. Most of these relations were developed by using laboratory data and sometimes they were verified by experiments at prototype scale. Development of a probabilistic approach to pier scour engineering requires a best-fit model rather than a conservative model. A study of the accuracy of the predictors can be carried out by comparing measured and predicted values of $y_{m,e}$. About 1000 experiments concerning scour around circular bridge piers were used to verify the relations discussed (Hoffmans, 1995). Johnson (1995) compared scour relations applying to 515 prototype experiments with both live-bed and clear-water scour. Johnson (1995) showed that the Breusers et al. (1977) relation appeared to work best as long as U_0/U_c remains above 0.5. The prediction accuracy of Equation (8.11). is about 78% for a discrepancy ratio r between the predicted and measured value in the range of 0.75–1.33, implying a standard deviation of 0.25μ.

8.5 Effects of specific parameters

8.5.1 Introduction

Many parameters affect the flow pattern and the process of local scour around bridge piers (Table 8.1). The influence of the fluid properties on the scour process is generally much less important than that of the flow characteristics. In addition, the size of bed material in the sand size range has little effect on scour depth. Larger-size bed material that will be moved by the flow will not affect the equilibrium scour depth, but will affect the time it takes to reach equilibrium. However, some laboratory studies indicate

Table 8.1 Scouring parameters

Fluid	g acceleration of gravity (m/s²) ρ fluid density (kg/m³) v kinematic viscosity (m²/s)
Flow	h_0 flow depth (m) U_0 mean flow velocity (m/s) C Chézy coefficient (m^{1/2}/s)
Bed material	• Representative diameter of the bed material: d_{50}, d_{84}, d_{16} • Gradation of the bed material, which determines the formation of an armour layer • Surface packing • Multiple layers of different bed materials • Grain form • Cohesion of material
Geometry	Shape in a horizontal section: • Circular, squared, rectangular, elliptic and other shapes Shape in a vertical section: • In elevation a constant shape (prismatic) • In elevation tapered structures or piers with an increased width at the base. Piers composed of different elements: • Pier with a ring-like footing to prevent scouring • Pier based on a foundation cap (footing) supported by foundation piles • Caisson with a flat foundation based directly on the subsoil • Caisson without a foundation, penetrated into the subsoil A group of separate piers, without a footing

that larger size particles in the bed material armour the scour hole and decrease scour depths (e.g. Breusers & Raudkivi, 1991). The width of the structure has a direct effect on the depth of scour, provided $b/h_0 < 1$. An increase in b increases scour depth. For slender bridge piers, the influence of the flow depth is marginal. The flow depth only has a direct effect on scour depth if $b/h_0 > 1$.

In rivers, increased bed roughness due to bed forms determines the flow depth and the form of the vertical velocity profile. The velocity of the approach flow influences the scour process. The higher the velocity, the deeper the scour depth. For live-bed conditions, the scour depth does not increase significantly with velocity. In such cases, there is a dynamic equilibrium between transport entering and leaving the scour hole.

Most studies have focused on local scour near circular prismatic piers in a non-cohesive sand bed, caused by a steady flow. This situation is a reference case for other geometries. The length of the pier has no appreciable effect on the scour depth as long as the pier is aligned with the flow. If the pier (e.g. a rectangular pier) is at an angle to the flow, the length influences the scour process significantly (Laursen & Toch, 1956). The foundation of a prismatic pier is normally a continuation of the pier for at least a few pier diameters into the subsoil. If the pier has a large diameter compared to its height, it can be regarded as a caisson.

The case of a single cylindrical pier has been used as a reference for all other cases and the deviations with respect to this simple case are generally expressed in generally applicable K factors:

$$K_i = K_s K_\omega K_g K_{gr} \tag{8.22}$$

where

K_g = factor for the influence of gradation of the bed material (-)
K_{gr} = factor for the influence of a group of piers (-)
K_i = correction factor (-)
K_s = pier shape factor (-)
K_ω = factor for orientation of the pier to the flow (-)

8.5.2 Pier shape

The influence of the shape of a horizontal cross-section of the pier has been investigated by Laursen and Toch (1956), Neill (1973) and Dietz (1972). Since their results confirm each other, recommended values of K_s have been set, as given in (Table 8.2). The shape of a vertical cross-section can also be expressed in a correction factor. For a basic case, i.e. a pier with a prismatic shape, K_s equals 1.0. Values of K_s are summarised in Table 8.2 for piers with a circular horizontal section and tapered elevation. No data for the shape factor for piers with a foundation footing or a pile-cap were found in the references.

8.5.3 Alignment of the pier to the flow

For cylindrical piers, it is not necessary to consider the influence of the angle of flow attack (i.e. the angle between the main axis of the horizontal cross-section of the pier and the direction of the approach flow) on the maximum equilibrium scour depth if the spacing between the piers is more than $3b$ to $11b$.

Table 8.2 Shape factor K_s

Form of cross-section	K_s
Horizontal shape	
Lenticular	0.7–0.8
Elliptic	0.6–0.8
Circular	1.0
Rectangular	1.0–1.2
Rectangular with semi-circular nose	0.90
Rectangular with chamfered corners	1.01
Rectangular nose with wedge-shaped tail	0.86
Rectangular with sharp nose 1:2–1:4	0.65–0.76
Vertical shape	
Pyramid-like (narrowing upwards)	0.76
Inverted pyramid (broadening upwards)	1.2

Laursen and Toch (1956) studied this influence for piers with a rectangular horizontal cross-section (Figure 8.10). If the angle of attack is zero, then $K_\omega = 1$. Nakagawa and Suzuki (1976) presented some results of scour around a rectangular cross-section with L_p/b about 1.75. The angle of attack varied from 0 to 45°. They found that the final scour for 30° is almost equal to that for 45° but much larger than that for zero degrees: $K_\omega = 1.3$–1.8. Froehlich (1988) gives the following relation for K_ω which agrees reasonably with the results given in Figure 8.10:

$$K_\omega = \left(\cos\omega + \left(L_p/b \right) \sin\omega \right)^{0.62} \qquad (8.23)$$

where
   $L_p =$ length of the pier (m)
   $\omega =$ angle of attack

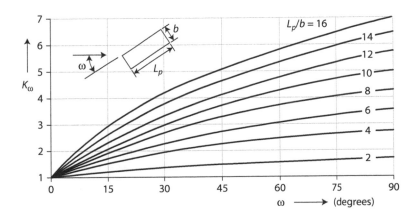

Figure 8.10 The influence of the alignment of the pier to the flow.

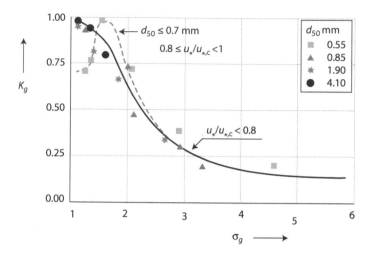

Figure 8.11 Factor K_g as a function of σ_g.

Note: K_ω should only be applied when the entire length is subjected to the attack of flow. When ω approaches 90°, results from Figure 8.11 and Equation (8.19) become questionable because this angle of flow attack essentially changes the shape of the pier.

8.5.4 Gradation of bed material

The influences of the grain diameter and the density of the material are usually expressed in the critical flow velocity for the initiation of sediment motion. The gradation of bed material can be characterised by the geometric standard deviation σ_g ($= d_{84}/d_{50}$), Vanoni (1977). For natural river sand σ_g is about 1.8, for uniform sand σ_g is 1.3 and for single grain size this ratio equals 1. In the basic case, the bed material consists of uniform sand with $K_g = 1$. The effect of the grain size distribution on the scour depth was reported by Raudkivi and Ettema (1985). Although little field data are available, we can conclude that increased grading of the bed material decreases the scour, thanks to the coarse particles in the bed material. The relation between K_g and σ_g in Figure 8.11 should be used with care.

8.5.5 Group of piers

Scouring at single piers has been more thoroughly investigated than scouring at groups of piers. In most cases, a reduction of scour is obtained when a solid pier is divided into smaller piers (e.g. Breusers et al., 1977). However, this reduction depends strongly on the pier spacing. The scour depth generally increases if the ratio between the pier spacing and the bridge pier width increases from about 1:4 to 1:8 (e.g. Herbich, 1984) because the individual piers cooperate as a single larger pier. Based on model experiments with a three-legged platform, Breusers (1971) showed an increase in scour over that for an isolated pile if the spacing is less than three diameters.

Table 8.3 The factor K_{gr} for a group of 2 circular piers

Flow direction	Pier position	Pier spacing	Front pier	Rear pier
		1b	1.0	0.9
		2–3b	1.15	0.9
		>15b	1.0	0.8
		1b	1.9	1.9
		5b	1.15	1.2
		>8b	1.0	1.0
		1b	1.9	1.9
		2–3b	1.2	1.2
		>8b	1.0	1.0

Breusers and Raudkivi (1991) describe mechanisms which affect scour at pile groups. Table 8.3 summarizes some of these results for a group of only two piers with a circular cross-section.

1. Two piers in a line parallel to the flow direction: The maximum scour depth around the front pier will increase by a maximum of 15% if the pier spacing is 2 to 3b. The influence of the second pier on the front pier scour disappears if the pier spacing is greater than 15b. The maximum scour depth of the rear pier is reduced by 10–20%. This reduction is almost independent of the pier spacing.
2. Two piers in a line perpendicular to the flow direction: The mutual influence of both piers on the maximum scour depth disappears if the pier spacing is greater than 10b.
   If the pier spacing is very small (say < 1.5b), both piers have almost the same effect as a single pier with a diameter of *2b*.
3. Two piers with a variable angle ω between the line through the piers and the flow direction (angle of attack): If the pier spacing is *5b or larger*, the maximum scour depth around the upstream pier (the front pier) is insensitive to the angle of attack. At an angle of attack of 45°, the maximum increase for the front and rear pier is approximately 15% and 20%, respectively.
   The results for two piers confirm the general rule that the resulting scour at a group of piers can be considered as the coincidental positioning of the separate scour holes of the individual piers.

8.6 Scour slopes

8.6.1 Introduction

According to Dargahi (1987), the slope of a scour hole around a bridge pier can be divided into three regions: upper and lower regions and, in the deepest part of a scour hole, a concave region (Figure 8.13). General methods used to predict scour slopes are treated in Sections 3.5.5 and 6.4. In the following, some specific values for bridge piers are mentioned. The shape, the slopes of a scour hole and the maximum scour depth determine the volume of a scour hole and the minimum extent of a bed protection to prevent local scour. First, the shape of a local scour hole around a cylindrical pier is treated and next the local scour hole around all other types of piers.

8.6.2 Single cylindrical pier

The reference case of a single pier in a uniform flow with its foundation in a uniform bed material has been studied in detail. For example, Dargahi (1987) describes in detail the development of the side slopes of a local scour around a single cylindrical pier during the scour process. He schematised the horizontal cross-sectional shape of the local scour hole into a half circle at the upstream side, connected to the half of an elongated ellipse at the downstream side. Bonasoundas (1973) used this schematisation for the bed protection around a cylindrical pier. The upstream part of the scour hole has different angles of slope for the upper and deeper parts of the scour hole (Figure 8.12). The slope in the deepest part of the scour hole is governed by the horseshoe vortex (see Figure 8.13). For reasons of simplicity, these two slope angles are averaged into one side slope. The upstream scour slope β_u is about equal to the angle of the natural slope ($\approx$ angle of repose ϕ). The average downstream scour slope β_d is generally less steep than the upstream scour slope and can be estimated by $\beta_d \approx 1/2\beta_u$.

In principle, the schematised shape of a local scour hole is a function of the side slopes, the radius of the upstream half circle and the length of the axis of the downstream half ellipse. The axes of a local scour hole are measured from the centre of the cylindrical pier. Upstream of the pier, the radius of a circle is about 2.5b, provided the flow depth is larger than three times the pier width. Downstream of the pier the shortest and longest axis of an ellipse amounts to 5b (Figure 8.14). With these slopes and widths of the schematised scour hole, the volume and the minimum extent of a bed protection can be determined. It should be noted that no difference has been observed between the shapes of the local scour holes in clear-water scour and that, as a result of segregation in the scour hole, the coarse fraction will form a layer on the bed of the scour hole.

For example, for slender bridge piers (circular and square) the equilibrium scour depth is 1.5 times the pier diameter. Assuming side slopes of 1V:2H, the width of the scour hole equals 7b (= 1b + (2 × 3)b). The downstream slope is less steep. If we assume a steepness of 1V:3H the length of the scour hole is 8.5b (= 1b + 3b + 4.5b). This results in scour hole dimensions as recommended by Bonasoundas (1973) (see Figure 8.14). The method of Hjorth (1975) is not recommended since it has been based on flume experiments only and has not been validated with prototype data.

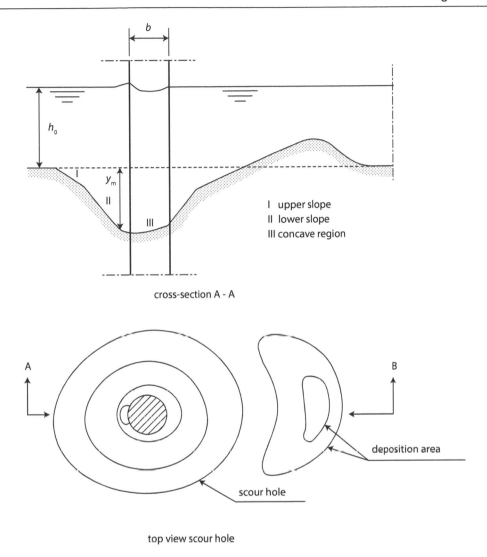

cross-section A - A

top view scour hole

Figure 8.12 **Definition sketch of scour around bridge piers.**

If the shape of the bridge pier is rectangular or elliptical, the scour dimensions depend strongly on the angle of attack of the flow. In such cases expert judgement is required to determine the bed protection, and experimental research is recommended. More information about dimensions of scour holes can be found in the literature.

8.6.3 Other types of piers

We have not found any systematic data about the shape of local scour holes around non-cylindrical piers in a non-uniform flow. However, the triaxial test for sand and the direct shear test for clay and peat give information about the angle of internal friction and this determines the steepness of the slopes. For example, for sand the angle of internal friction varies between 30 and 40 degrees.

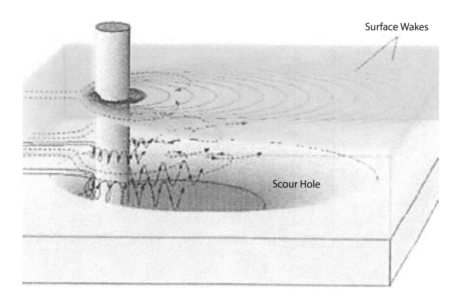

Figure 8.13 **Three-dimensional sketch of a scour hole at a bridge pier.**

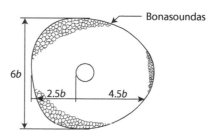

Recommended scour protection

Figure 8.14 **Recommended riprap protection.**

We obtained average values for the slope steepness by combining data from different sources. We measured characteristic values of the side slopes of local scour holes from graphs and figures in different publications. None of these publications included any information about the angle of repose, so the angle has been estimated. For the estimates we assumed that in the laboratory tests the sand had a loose compaction and that in prototype situations, after many years of consolidation, the soil had a firm compaction. In some situations, the flow depth is less than $3b$. Although the flow depth reduces the maximum scour depth to some extent, we assumed that this does not affect the shape of the scour hole. Despite the lack of systematic data, it might be concluded that the shape of the pier probably has no influence on the maximum upstream side slope of the local scour hole.

In an estuary with tidal flow, we defined in the direction of the flow which sides of the pier are upstream and downstream relative to the highest flow velocity during ebb or flood flow. Furthermore, the side slopes of the scour hole being normal to the flow direction might be vulnerable for slidings and flow slides.

These conclusions are based on a limited number of publications and need further confirmation by other studies. This also holds for the assumption that the influence of flow depth on scour hole shape is negligible. The influence of gradation of the subsoil and the flow direction relative to the pier axis on this side slope of a local scour hole cannot be assessed from this limited number of publications.

Sometimes, bridge piers are placed on a caisson or a footing (see Figure 8.15). HEC-18 and CIRIA report C742 (2015) provide methods to compute the scour for these situations. Note that in the Netherlands the bed around a bridge pier is protected, whereas this is not common in e.g. the US and New Zealand.

The scour can be computed with the formulas presented in Section 8.4.2, in particular the Hoffmans formula. For particular situations the reader is referred to HEC-18 or CIRIA C742.

If a footing or design cap is applied in erodible bed material, it is recommended to keep the top level of the footing or pile cap below the lowest level of the natural bed + contraction scour (see left sketch in Figure 8.15). In this situation, the footing or pile cap can be considered as a solid, non-erodible layer, and it may reduce the scour at least in the initial phase of scouring. If scour develops, extra turbulence around the footing will make the scour hole larger and deeper. Thus, if scour develops below the top level of the footing it is crucial in the design phase to estimate the level of the footing relative to the natural bed level + contraction scour.

If in a final design the top of the footing is already above the lowest level of the natural bed level + contraction scour, see right sketch in Figure 8.15, then the footing or pile cap will create a larger obstacle for the flow resulting in more scour. Then, the scour depth will be determined by the width of the footing or the pile cap.

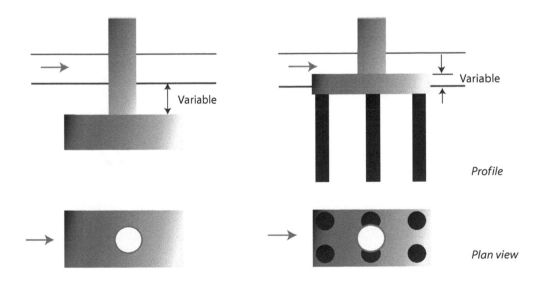

Figure 8.15 **Bridge pier on a footing (left) and a pile cap (right) (CIRIA, 2015).**

If the footing is higher than the bed, the approach width of the footing (and if relevant including the angle of approach of the flow) determines the scour instead of the pier diameter (see left below sketch in Figure 8.15).

8.6.4 Winnowing

In general, the bed protection applied consists of an armour layer on a filter layer which does not allow the transport of fines of the base material, the so-called geometrically closed filters. However, it is also possible to apply a geometrically open filter that reduces the hydraulic load near the base layer to limit or prevent transport of base material through the filter layer. The entrainment of fine particles (base material) in the armour layer is called winnowing. Laboratory experiments were conducted to investigate winnowing at circular piers in relation to a varying current velocity, filter thickness and stone size.

Sonneville et al. (2014) summarised the setup, test programme, monitoring techniques and results of the conducted experiments. The results of the experiments show that the winnowing process is characterised by a gradual subsidence of the filter stones into the base layer. The winnowing depth is defined as the scour adjacent to a pier and related to an open filter which allows limited loss of base material. The scour increases with current velocity and decreases with a thicker filter layer. A design rule was derived to estimate the winnowing depth for a single protection layer with a relative thickness D_F/d_{50f} based on a given flow $\Omega_c = \Psi/\Psi_c$ (see dashed lines in Figure 8.16):

$$\frac{y_{m,e}}{b} = \exp\left(-\frac{4.5}{\Omega_c}\cdot\left(D_F\big/d_{50f} -1.5\right)^{0.66}\right) \quad \text{for} \quad D_F\big/d_{50f} > 1.5 \tag{8.24}$$

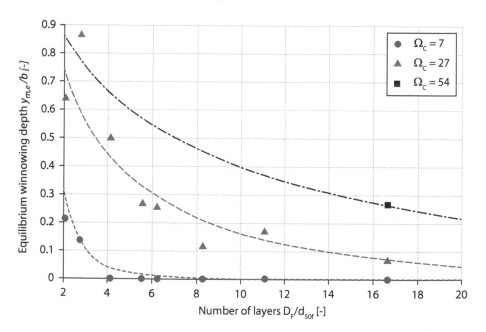

Figure 8.16 Equilibrium winnowing depth vs. filter thicknesses for different hydraulic loads.

where:

Ω_c = current-related sediment mobility,

$\Omega_c = \Psi/\Psi_c$ (Whitehouse, 1998),

Ψ = is the Shields parameter and Ψ_c is the critical Shields parameter.

D_F = filter thickness

d_{50f} = median filter size

The required filter thickness D_F can be determined with report CUR233 (2010).

Note: the equation has not been validated with prototype experiments and is thus not generally applicable.

8.7 Measures to prevent local scour

8.7.1 Introduction

The maximum scour depth of a scour hole is critical for failure of a bridge pier. The volume and the slopes of a scour hole have no direct influence on bridge pier stability unless the slope becomes too steep. This section describes measures to prevent or to reduce local scour around a bridge pier. Countermeasures can be distinguished in the following types (for more details, see Section 2.4):

1. Slope-hardening countermeasures
2. Slope-stabilising countermeasures
3. Flow-altering countermeasures

8.7.2 Riprap protection

The size of the riprap can be calculated by using the stability criteria of Izbash or Shields, if the maximum critical flow velocity is known. The maximum shear stress (τ_{max}) and the maximum flow velocity (U_{max}) near a cylinder type pier can be calculated:

$$\tau_{max} = 4\tau_{undisturbed} \tag{8.25}$$

and

$$U_{max} = 2U_{undisturbed} \tag{8.26}$$

Equations (8.25) and (8.26) compute an upper limit and are based on potential flow. The minimum layer thickness is $2d_{50}$, but often a greater thickness will be advantageous. In particular, the connection between the riprap layer and the footing of the pier needs extra attention. Some qualitative information is given in Ettema (1980).

Stone bags around a pier provide a type of protection that guarantees a flexible connection of the riprap with the pier. Reasonable results have been obtained with this type of protection at the Eastern Scheldt Storm Surge Barrier and the Ekofisk offshore structure. Grouted gravel or sand has been used to repair damage caused by scour under the bed of a caisson. Some preliminary design rules for the length of the protection layer L (or the diameter of the protection layer) are available from the literature.

Bonasoundas (1973) relates the minimum length of the bed protection around circular bridge piers to the projected width of the pier b. The length of the bed protection is about $2.5b$ upstream, and downstream of bridge piers the critical failure length is $L_s = 4b$. If the width of the pier influences the flow pattern strongly (e.g. b is half the width of the flow), the relation given by Bonasoundas no longer holds.

Carstens (1976) found that L_s is dependent on the maximum scour depth y_m and can be given by:

$$L_s = \gamma_s y_m \cot \phi \qquad (8.27)$$

where γ_s = safety factor (–)

Equation (8.27) computes an upper limit because of the presence of the safety factor.

The angle of repose depends on soil type and compaction. For non-cohesive sediments, ϕ ranges from 30° for fine sand to 40° for coarse sand, and from 40° to 45° for gravel and rock. Assuming $\phi = 30°$ and $F_s = 1.5$, a conservative value for L_s is obtained ($L_s = 2.6y_m$). Zanke (1994) proposed a self-filling riprap protection system using a reservoir in the pile (Figure 8.14). A physical model investigation showed that this concept gave promising results.

Note: The Carstens equation has a relation with the failure length and shear flow (see Sections 3.5.5 and 3.5.6).

8.7.3 Mattress protection

For local protection around a big, circular pier in a bed of fine sand, an artificial protection has been proposed. This protection consists of numerous bundles of polyester filaments which can be suspended under a frame cantilevered from the pier. The first prototype test showed promising results (Carstens, 1976). In general, special attention has to be given to providing a tight connection between the mattress and the pier because even through a small gap, the downflow could induce severe erosion that extends under the mattress. The concept of using mattresses was put into practice at the Eastern Scheldt Barrier.

Recently, Hawkswood and King (2016) published possible failure modes for different types of mattresses (see Table 8.4).

Table 8.4 **General stability principal: Scour protection fails when the hydrodynamic loads upon it exceeds the stabilising resistance**

Type	Principle failure mode		
Rock/In-situ Concrete	Particle displacement	(1)	
	Uplift pale failure	(2)	
• Concrete mattress	Edge underscour	(3)	
• Concrete slabs			
Prefabricated Mattress	Joint failures		
	Unit movement	(3)	
• Concrete block mattress	Edge underscour		
• Gabion/reno mattress	Others		

8.7.4 Deflectors

The intensity of the downflow near the pier can be reduced by a deflector (e.g. Carstens, 1976; Dargahi, 1987). No design rules are available for the determination of the width and the height of the deflector, which is fixed to the pier. However, Chabert and Engeldinger (1956) found that the scour depth can be reduced by 60% by using a collar with a diameter of $3b$ located at a depth of $0.4b$ under the original bed level. Carstens published some results for the height of the deflector in only one situation. Dargahi investigated the shape of the collar and the position of the collar relative to the original bed level. The shape varied from circular collars to so-called 'Joukowski collars' (egg-shaped). No significant difference was found between the maximum scour depths for these two shapes. A collar at or just below the original bed level gives the highest reduction of the maximum scour depth and a slower development of the scour hole. For a preliminary design of the piers of the Eastern Scheldt Storm Surge Barrier, Akkerman (1976) studied the effectiveness of a horizontal separated deflector around a prismatic pier in a scale model. With an open space of $0.3\,\mathrm{m}$ between the deflector and the pier, the maximum scour depth was reduced by about 25%. As an alternative in this preliminary design, Akkerman (1976) also tested the effectiveness of a separated vertical deflector around a prismatic pier in a scale model. This deflector was placed on the bed protection mattress. The distance between deflector and pier was around $4\,\mathrm{m}$ and the gap between the mattress and the pier was around $0.3\,\mathrm{m}$. Other data are $b = 18\,\mathrm{m}$ and $h_0 = 15\,\mathrm{m}$. This deflector did not noticeably reduce the erosion depth.

8.8 Example

8.8.1 Introduction

An example is given about the local scour around bridge piers. The objective is to show the differences between slender and wide piers but also the effect of the turbulence intensity.

8.8.2 Local scour around bridge piers

Two different configurations are considered: narrow slender piers and wide piers, each with their own distinctly different flow field characteristics. The definition of narrow or wide in this respect depends on the ratio between the width b of the pier and the flow depth h_0. For slender piers, this ratio is less than 1 and for wide piers more than 1 (see Figure 8.17).

For each configuration, various equations have been used and the results are presented in tables and figures showing the scour depth as a function of the relative turbulence. The used equations are:

$$y_{m,e} = 1.4b \qquad (8.5)$$

$$y_{m,e} = 1.5 K_i b \tanh\left(h_0/b\right) \qquad (8.6)$$

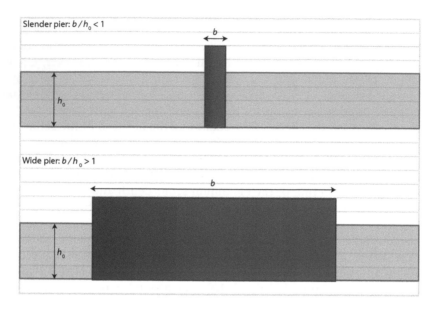

Figure 8.17 **Examined pier configurations.**

$$y_{m,e} = 2.0 K_i h_0 \mathrm{Fr}^{0.43} \left(\frac{b}{h_0} \right)^{0.65} \tag{8.9}$$

$$\frac{y_{m,e}}{b} = 1.6\chi_e \left(\frac{U_0}{U_c} \right)^2 - 1.3 \quad \text{for} \quad U_0 < U_c \tag{8.11}$$

$$\frac{y_{m,e}}{b} = 1.6\chi_e - 1.3 \quad \text{for} \quad U_0 \geq U_c \tag{8.12}$$

$$\chi_e = \frac{1 + 6.3 r_0^2}{1 - 6.3 r_{0,m}^2} \tag{8.13}$$

8.8.2.1 Slender piers

First, the effect of the relative turbulence intensities r_0 and $r_{0,m}$ will be shown with the Hoffmans Equations (8.11), (8.12) and (8.13) for clear water and live bed scour. The results are presented in Tables 8.5 and 8.6 respectively and in Figure 8.18. Clearly, can be seen that the equilibirum scour depth increases with an increasing value of $r_{0,m}$ and obviously, also with an increasing value of r_0. The results also show that the equilibirum scour depth is larger for live bed scour than for clear water scour, and can reach depths upto almost $3m$ for values of $r_{0,m} = 0.30$.

In order to show the differences between the older Equations (8.5) and (8.6) of Breusers and the CSU formula Equation (8.9) the scour depths are computed in Tables 8.7 up to 8.9. The computed results are depicted in Figure 8.19. Clearly can be seen that for water depths larger than 3m the scour depths hardly differ with values of 1.4 to 1.6m.

Table 8.5 Clear water scour with Equations 8.11 and 8.13 for $r_0 = 0.1$ and $r_0 = 0.15$

b (m)	r_0 (-)	$r_{0,m}$ (-)	U_0 (m/s)	U_c (m/s)	χ_e (-)	$y_{m,e}$ (m)
1	0.1	0.15	1	1.1	1.2	0.3
1	0.1	0.20	1	1.1	1.4	0.6
1	0.1	0.25	1	1.1	1.8	1.0
1	0.1	0.30	1	1.1	2.5	1.9
1	0.15	0.15	1	1.1	1.3	0.5
1	0.15	0.20	1	1.1	1.5	0.7
1	0.15	0.25	1	1.1	1.9	1.2
1	0.15	0.30	1	1.1	2.6	2.2

Table 8.6 Live bed scour with Equations 8.12 and 8.13 for $r_0 = 0.1$ and $r_0 = 0.15$

b (m)	r_0 (-)	$r_{0,m}$ (-)	χ_e (-)	$y_{m,e}$ (m)
1	0.1	0.15	1.2	0.7
1	0.1	0.20	1.4	1.0
1	0.1	0.25	1.8	1.5
1	0.1	0.30	2.5	2.6
1	0.15	0.15	1.3	0.8
1	0.15	0.20	1.5	1.1
1	0.15	0.25	1.9	1.7
1	0.15	0.30	2.6	2.9

Note that these scour depths are much smaller than the computed scour depths with the new Breusers Equations (8.11) to (8.13) for values of $r_{0,m}$ larger than 0.25. For smaller values the older formulas result in too conservative estimates of the scour depths.

8.8.2.2 Wide piers

Similar as shown for slender piers, in this section the effect of the relative turbulence intensities r_0 and $r_{0,m}$ will be shown with the Hoffmans Equations (8.11), (8.12) and (8.13)

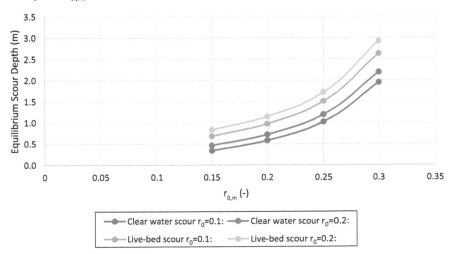

Figure 8.18 Sensitivity scour depth for turbulence intensity for slender piers.

Table 8.7 Live bed scour with Equation 8.5

b (m)	h_0 (-)	$y_{m,e}$ (m)
1	1	1.4
1	3	1.4
1	5	1.4
1	7	1.4

Table 8.8 Live bed scour with Equation 8.6

b (m)	h_0 (-)	K_i (-)	$y_{m,e}$ (m)
1	1	1	1.1
1	3	1	1.5
1	5	1	1.5
1	7	1	1.5

Table 8.9 Live bed scour with Equation 8.6

b (m)	h_0 (-)	K_i (-)	U_0 (m/s)	Fr (-)	$y_{m,e}$ (m)
1	1	1	1	0.32	1.2
1	3	1	1	0.18	1.4
1	5	1	1	0.14	1.5
1	7	1	1	0.12	1.6

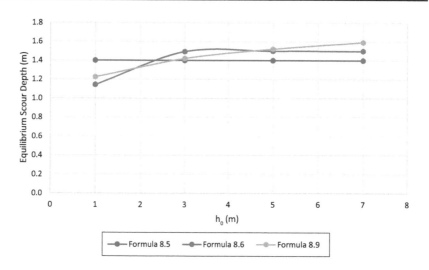

Figure 8.19 Influence of water depth on scour depth for slender piers

for clear water and live bed scour. Note that in Equations (8.11) and (8.12) the same values for the coefficients have been used while in Section 3.4.3 is stated that they are unknown for wide piers. The results are presented in Tables 8.10 and 8.11 respectively and in Figure 8.20. The results show the same tendencies as for slender piers: larger equilibirum scour depths with increasing values of $r_{0,m}$ and r_0, and also larger equilibirum

Table 8.10 Clear water scour with Equations 8.11 and 8.13 for $r_0 = 0.1$ and $r_0 = 0.15$

b (m)	r_0 (-)	$r_{0,m}$ (-)	U_0 (m/s)	U_c (m/s)	χ_e (-)	$y_{m,e}$ (m)
7	0.1	0.15	1	1.1	1.2	2.4
7	0.1	0.20	1	1.1	1.4	4.1
7	0.1	0.25	1	1.1	1.8	7.1
7	0.1	0.3	1	1.1	2.5	13.6
7	0.15	0.15	1	1.1	1.3	3.2
7	0.15	0.20	1	1.1	1.5	5.0
7	0.15	0.25	1	1.1	1.9	8.3
7	0.15	0.3	1	1.1	2.6	15.3

Table 8.11 Live bed scour with Equations 8.12 and 8.13 for $r_0 = 0.1$ and $r_0 = 0.15$

b (m)	r_0 (-)	$r_{0,m}$ (-)	χ_e (-)	$y_{m,e}$ (m)
7	0.1	0.15	1.2	4.8
7	0.1	0.20	1.4	6.8
7	0.1	0.25	1.8	10.5
7	0.1	0.3	2.5	18.4
7	0.15	0.15	1.3	5.8
7	0.15	0.20	1.5	8.0
7	0.15	0.25	1.9	12.0
7	0.15	0.3	2.6	20.4

scour depth for live bed scour than for clear water scour. Obviously, the scour depths are much larger than for slender piers due to the bridge pier width of 7 m, reaching values of 20m for values of $r_{0,m} = 0.30$.

Also for wide piers the differences between the older Equations (8.5) and (8.6) of Breusers and the CSU formula Equation (8.9) the scour depths will be shown, see the Tables 8.12 up to 8.14 and Figure 8.21. For water depths smaller than 7m the scour

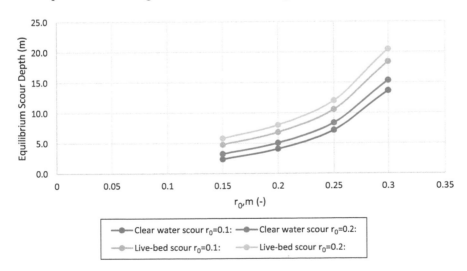

Figure 8.20 Sensitivity scour depth for turbulence intensity for wide piers.

Table 8.12 Live-bed scour with Equation 8.5

b (m)	h_0 (-)	$y_{m,e}$ (m)
7	1	9.8
7	3	9.8
7	5	9.8
7	7	9.8

Table 8.13 Live bed scour with Equation 8.6

b (m)	h_0 (-)	K_i (-)	$y_{m,e}$ (m)
7	1	1	1.5
7	3	1	4.2
7	5	1	6.4
7	7	1	8.0

Table 8.14 Live bed scour with Equation 8.9

b (m)	h_0 (-)	K_i (-)	U_0 (m/s)	Fr (-)	$y_{m,e}$ (m)
7	1	1	1	0.32	4.6
7	3	1	1	0.18	5.0
7	5	1	1	0.14	5.4
7	7	1	1	0.12	5.6

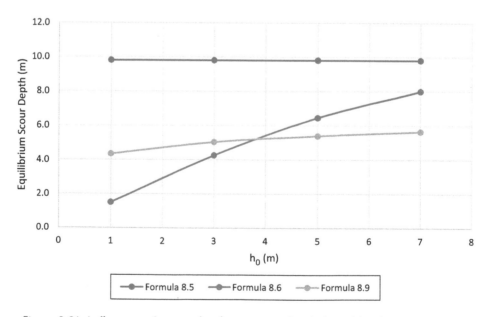

Figure 8.21 Influence of water depth on scour depth for wide piers.

depths are comparable with the ones computed for relative turbulence intensities of maximum 0.25. Both Breusers formulas predict for water depths larger than $7m$ a scour depth around $10m$; whereas the CSU formula tends to a maximum of 6m. These values are much smaller than for high relative turbulence intensities of $r_{0,m}$ of 0.3. It can be concluded that it is important to know the correct value of r0,m. In Section3.4.3 Equation (3.15) is given and as a first estimate a value of 0.25.

Chapter 9

Case studies on prototype scale

9.1 Introduction

Nine case studies of prototype projects are presented to show the application of the design formulas presented in the manual. In some case studies, scour results of numerical models are compared with predicted values with the design formulas, while in other case studies computations with analytical formulas are compared with observed scour. In general, we stress the need of checking numerical results with analytical formulas. In many of the cases, a bed protection is needed to prevent scouring close to the hydraulic structure for which the design is discussed.

The case studies address different hydraulic structures, viz. bridge piers, culverts, sills and ship propellers. The case studies have been provided by consultants, contractors, research institutes and Rijkswaterstaat, who are responsible for the content. The size and the quality of the cases differ, but all contributions show the approach of the authors to come up with an appropriate design with the tools available. The case studies are:

Section	Case study	Scour type
9.2	Camden motorway bypass bridge pier scour assessment	Bridge pier scour
9.3	Project Waterdunen	Culvert scour
9.4	Full-scale erosion due to main ship propeller	Jet scour
9.5	Scour due to ship thrusters	Jet scour
9.6	Crossing of high voltage power line	Bridge pier scour
9.7	Scour development in front of a culvert	Culvert scour
9.8	Bed protection at the railway bridge in bypass of the River Waal	Bridge pier scour
9.9	Pressure scour around bridge piers	Bridge pier scour
9.10	Bed protection at the weir in the River Meuse at Grave	Sill scour

9.2 Camden motorway bypass bridge pier scour assessment (RHDHV)

9.2.1 Introduction

This case addresses constriction and local pier scour at a motorway bypass over a river and its floodplain. The flow and scour are estimated using numerical models for different flow conditions. After the introduction, Sections 9.2.2–9.2.6 present the

Figure 9.1 Camden bypass motorway passing over the Nepean River floodplain.

applied method, formulas and results using the HEC-RAS model. In Section 9.2.7, the numerical results are verified with analytical methods, as presented in Chapter 8. In Section 9.2.8, the resulting scour protection is discussed. Section 9.2.9 presents the conclusions.

In 2014, an application was lodged for concent to excavate soil and sand from the floodplain of the Nepean River, upstream of the Camden Bypass. The bypass is located in the southwest of Sydney, Australia and conveys vehicles over the Nepean River and floodplain via a suspended bridge (Camden Bypass Bridge). The bridge is founded on a series of dual piers (see Figure 9.1).

The proposed excavation was located approximately 100 m upstream of the bridge and comprises a 40,000 m^2 area up to 4 m deep. In the event of extreme flooding, the floodplain under the bridge is inundated by flood water. Excavation upstream of the Camden Bypass Bridge was estimated to increase floodplain flow velocity downstream of the excavation (at the bridge). Subsequently, consideration was given to the impact of the pit extension on hydraulics and the associated potential scour around bridge piers.

As the pit extension was estimated to impact on the peak velocities of flood water under the bridge, an assessment of potential impacts on scour was requested. A scour analysis was undertaken for both the existing situation and a scenario following the proposed excavation. Both constriction scour and local pier scour were considered. Total scour (constriction plus local scour) was estimated at each of the 25 piers.

9.2.2 Assessment of scour

The impact of the pit extension on hydraulics was estimated using the TUFLOW 2D hydraulic model. The impact of the pit extension on peak velocity was assessed for a number of flood events, namely 20-, 100- and 2000-year Average Recurrence

Interval (ARI). The 2000-year ARI is generally considered an upper limit for bridge assessments.

The proposed pit extension was estimated to have the greatest impact on peak velocities downstream of the pit (under the bridge). Greatest estimated impacts on the peak flow velocity (i.e. the difference between the existing situation and the proposed excavation scenario) were found to occur during the 20-year ARI event, with a maximum increase in peak velocity of 0.94 m/s under the bridge, compared to a maximum increase of 0.40 m/s and 0.54 m/s in the 100- and 2000-year ARI events, respectively. It should be noted that, although the 20-year ARI event produces the greatest impacts on peak velocity between the existing situation and proposed excavation scenario, the 100- and 2000-year ARI events produce greater depths of scour, in both the existing situation and the proposed excavation scenario.

A bridge scour analysis was undertaken using the one-dimensional (1D) hydraulic model HEC-RAS. This software package was selected for its ability to model bridges and the resultant contraction, abutment and pier scour associated with floodwaters. The model was developed using a steady-state condition as the peak flow is of main concern for bridge scour, due to the velocity of floodwater at the flood peak.

The baseline hydraulics of the HEC-RAS model were calibrated to the results of an existing 2D TUFLOW hydrodynamic flood model (Figure 9.2). The 2D TUFLOW model provided the most accurate hydrodynamic results to feed into the HEC-RAS scour model.

The scour analysis was completed in HEC-RAS for both the 100-year and 2000-year ARI floods using the existing and simulated proposed conditions. The total scour

Figure 9.2 Mapped flood extent, depth and velocity distribution taken from the TUFLOW model for the 2000-year ARI flood event.

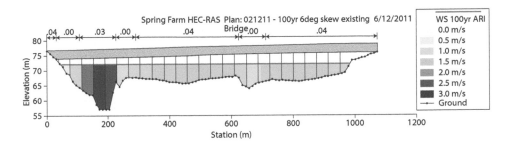

Figure 9.3 **Extract from ID HEC-RAS model showing peak velocity for the 100-year ARI flood event.**

depths at each bridge pier and abutment were determined. It should be noted that for the 100-year ARI event, only the bridge piers offer any resistance to flow as the 100-year ARI flood level is below the bridge abutments which are at the deck level (see Figure 9.3, which shows the cross-section at the bridge).

To determine scour depths, the model required the average soil particle diameter at which 50% (d_{50}) and 95% (d_{95}) of soil particles is smaller. These sizes were estimated for each pier and location along the floodplain cross-section using geotechnical investigation data. The local flood velocities and water depths for each type of scour were also utilised to estimate the scour profile under the bridge.

9.2.3 Scour assessment results

The results from the HEC-RAS scour modelling provide estimates of constriction, pier and abutment scour. Constriction scour occurs in the channel or overbanks if the bridge infrastructure constricts the floodwaters. Pier and abutment scour are caused by the dispersion of flow due to these obstructions. Figure 9.4 displays the total of all types of scour expected in the 100-year ARI flood for the existing site conditions.

9.2.4 Constriction scour

The results of the constriction scour analysis are summarised in Table 9.1. HEC-RAS uses Laursen's live-bed constriction-scour equations (Laursen, 1960) for this, as presented below:

$$\frac{h_2}{h_1} = \left(\frac{Q_2}{Q_1}\right)^{6/7} \left(\frac{W_2}{W_1}\right)^{k_1} \left(\frac{n_2}{n_1}\right)^{k_2} \tag{9.1}$$

The average scour depth follows with: $y_{m,e} = h_2 - h_0$ (m)
   where

$y_{m,e}$ = equilibrium scour depth (m)
h_2 = average depth in the upstream main channel after scour (m)
h_1 = average depth in the contracted section (m)

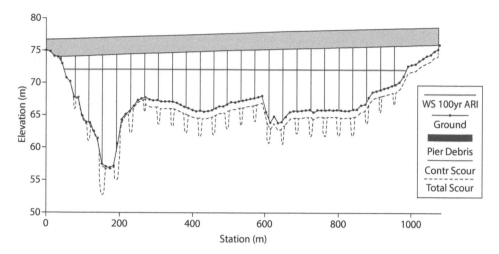

Figure 9.4 HEC-RAS output showing cross-section of bridge and piers, along with general and local scour estimates.

Table 9.1 Estimated constriction scour, showing maximum (equilibrium) scour and estimated increase in scour between the existing and proposed excavation scenario (negative values indicate a decrease in scour depth due to the proposed excavation)

Flood event	Location	Critical equilibrium scour (m)	Increase/decrease in scour (difference) between existing situation and proposed excavation scenario (m)
100-year ARI	Left overbank	0	0
	Channel	0.27	−0.27
	Right overbank	1.46	0.36
2000-year ARI	Left overbank	0.23	0.02
	Channel	1.19	0.04
	Right overbank	1.79	0.04

h_0 = existing depth in the contracted section before scour (m)
Q_1 = discharge in the upstream channel transporting sediment (m³/s)
Q_2 = discharge in the contracted channel (m³/s)
W_1 = bottom width of the upstream main channel (m)
W_2 = bottom width of the main channel in the contracted section (m)
n_1 = Manning's n for the upstream main channel (-)
n_2 = Manning's n for contracted section (-)

There is an estimated minor decrease in constriction scour in the main channel for the 100-year ARI flood. This is explainable as the flow is redistributed as a result of the proposed floodplain excavation, causing a slight reduction in velocities in some areas and subsequently less scour. The greatest estimated increase in scour due to the proposed excavation was found to be 0.36 m on the right overbank during a 100-year ARI flood event. The critical constriction scour was found to be in the right overbank at 1.79 m for the 2000-year recurrence flood.

9.2.5 Abutment scour

HEC-RAS uses Richardson's equation (Richardson et al., 1990) for abutment scour (also known as the CSU formula for bridge pier scour):

$$y_{m,e} = 2.0 K_i h_0 Fr^{0.43} \left(\frac{b}{h_0} \right)^{0.65}$$
(8.9)

with

$$Fr = U_0 / (gh_0)^{0.5}$$

where:
   K_i = correction factor for various influences (–); K_i = 1.2
   b = width of bridge pier (m)
   Fr = Froude number (–)
   U_0 = flow velocity (m/s)

No abutment scour was predicted at either end of the bridge for the 100-year ARI flood as the water level for this instance was sufficiently low to avoid significant redistribution. The 2000-year ARI flood, however, causes a maximum estimated abutment scour depth of 6.7 m and 3.9 m at the western and eastern abutments, respectively. The assessment of abutment scour shows a minor increase in the total scour of 0.5 m as a result of the excavation, compared to the existing situation.

9.2.6 Pier scour

HEC-RAS uses Richardson's equation (Richardson et al., 1990) for pier scour, as presented above. Table 9.2 summarises results of the pier scour analysis.

Table 9.2 Estimated critical pier scour as well as average and maximum estimated increases

Flood event	Maximum pier scour (m)	Average scour increase to maximum pier scour across all piers (m)	Maximum scour increase to maximum pier scour (m)
100-year ARI	4.80	0.46	0.95
2000-year ARI	5.85	0.27	0.39

Table 9.3 Estimated sum of constriction and pier scour depths for the existing and proposed excavation scenario for the 100-year and 2000-year ARI flood event and variance (difference) between the existing and proposed scenarios

Pier no. (as per technical drawings)	100-year existing scour (m)	100-year proposed scour (m)	Variance (m)	2000-year existing scour (m)	2000-year proposed scour (m)	Variance
1	3.07	3.38	0.31	4.23	4.47	0.24
2	3.43	3.74	0.31	4.42	4.68	0.26
3	3.58	3.88	0.3	4.5	4.77	0.27
4	3.99	4.28	0.29	4.76	5.04	0.28
5	4.01	4.3	0.29	4.77	5.05	0.28
6	4.01	4.3	0.29	4.77	5.05	0.28
7	4.02	4.2	0.18	4.78	5.06	0.28
8	3.5	3.62	0.12	4.22	4.46	0.24
9	3.57	3.68	0.11	4.27	4.51	0.24
10	3.7	4.65	0.95	4.48	4.73	0.25
11	3.61	4.24	0.63	4.69	4.97	0.28
12	3.76	4.39	0.63	4.78	5.06	0.28
13	3.79	4.43	0.64	4.8	5.09	0.29
14	3.98	4.64	0.66	4.92	5.22	0.3
15	4.02	4.68	0.66	4.95	5.25	0.3
16	3.98	4.64	0.66	4.93	5.22	0.29
17	3.78	4.42	0.64	4.79	5.08	0.29
18	3.6	4.19	0.59	4.6	4.87	0.27
19	2.19	1.88	0.31	3.26	3.48	0.22
20	3.59	3.25	−0.34	4.56	4.88	0.32
21	4.8	4.43	−0.37	5.46	5.85	0.39
22	4.67	4.3	−0.37	5.37	5.75	0.38
23	2.84	2.52	−0.32	3.75	4.01	0.26
24	1.77	1.73	−0.04	2.23	2.41	0.18
25	0	0	0	1.99	2.16	0.17

Table 9.3 presents the estimated individual pier scour and includes both the existing situation and the proposed scenarios for the 100-year and 2000-year ARI events, respectively. The 'variance' indicates the estimated difference between the scour depths in the existing situation and in case of the proposed excavation scenario.

9.2.7 Numerical Model Verification

To provide verification and a cross-check of the computational assessment as outlined above, calculations were undertaken using the design equations presented in this scour manual as well as in the CIRIA manual on scour at bridges and other hydraulic structures (CIRIA, 2002) (refer to Section 4.3.2 of that manual). The equations used for the comparison are presented below.

Based on the data of Laursen and Toch (1956), Breusers et al. (1977) found a relation for circular bridge piers:

$$y_{m,e} = 1.35 K_i b^{0.7} h_0^{0.3} \tag{8.4}$$

$$K_i = K_s K_\omega K_g K_{gr} \tag{8.22}$$

where:

K_g = factor for the influence of gradation of the bed material
K_{gr} = factor for the influence of a group of piers
K_s = pier shape factor
K_ω = factor for orientation of the pier to the flow

Equation (8.8) proposed an upper-limit formula (Breusers et al., 1977). For live-bed conditions, the equation becomes:

$$y_{m,e} = 2K_i b \tanh(h_0/b) \quad \text{for} \quad U_0/U_c > 1 \tag{8.8}$$

where:

U_c = critical mean velocity (m/s)

For circular bridge piers the Colorado State University's equation (Johnson, 1992) is as follows (the correction factor making the equation applicable to other pier shapes):

$$y_{m,e} = 2.0 K_i h_0 \text{Fr}^{0.43} \left(\frac{b}{h_0}\right)^{0.65} \tag{8.9}$$

Section 4.3.2 of the C551 CIRIA manual (CIRIA, 2002) presents the following modified version of the non-dimensional equilibrium-depth-of-scour equation originally developed by Breusers et al. (1977), along with details on its application:

$$\frac{y_{m,e}}{b} = \gamma_s \phi_{shape} \phi_{depth} \phi_{velocity} \phi_{angle} \tag{9.2}$$

where:

b = horizontal width of the structure measured normal to its longitudinal axis (m)
γ_s = factor of safety (–)
ϕ_{shape} = shape factor taking into account the effect of the shape of the structure (–)
ϕ_{depth} = depth factor taking into account the effect of relative water depth (–)
$\phi_{velocity}$ = velocity factor taking into account the effect of the flow velocity (–)
ϕ_{angle} = angle factor taking into account the effect of an angle in the approach flow relative to the longitudinal axis of the structure (–)

The results of the computational assessment using a numerical model have been compared to calculations based on abovementioned empirical equations. Table 9.4 presents results of the comparison. Calculations were generally comparable to the numerical modelling, but results did differ between the equations applied.

9.2.8 Scour mitigation

Following the scour estimates made above, a structural assessment was undertaken on the bridge to determine potential impacts and inform scour mitigation measures. Bridge piers 7, 13 and 19 were determined most significant to the bridge's structural

Table 9.4 100-year ARI event equilibrium scour depth results in existing situation based on numerical modelling and empirical equations

Pier	Computer calculations (m)	Hand calculations results (m)			
		Equation (9.2)	Equation (8.4)	Equation (8.8)	Equation (8.9)
1	3.07	3.23	2.76	3.28	2.14
2	3.43	3.95	2.99	3.47	2.64
3	3.58	4.28	3.44	3.59	2.81
4	3.99	4.32	3.72	3.60	2.91
5	4.01	4.32	3.72	3.60	2.91
6	4.01	4.32	3.74	3.60	2.92
7	4.02	4.32	3.70	3.60	2.91
8	3.50	4.32	3.79	3.60	1.83
9	3.57	4.32	4.00	3.60	3.01
10	3.70	4.28	3.89	3.60	2.97
11	3.61	4.32	3.33	3.58	2.77
12	3.76	4.32	3.44	3.59	2.81
13	3.79	4.32	3.59	3.60	2.87
14	3.98	4.32	3.70	3.60	2.91
15	4.02	4.32	3.77	3.60	2.93
16	3.98	4.32	3.65	3.60	2.89
17	3.78	4.32	3.47	3.59	2.82
18	3.6	4.28	3.44	3.59	2.36
19	2.19	4.28	3.38	3.58	1.74
20	3.59	4.32	3.74	3.60	3.63
21	4.80	4.32	4.53	3.60	4.29
22	4.67	4.32	4.80	3.60	3.70
23	2.84	4.32	4.05	3.60	2.54
24	1.77	3.60	3.31	3.57	1.72
25	0	0	0	0	0

integrity, as these piers were anchor piers. Anchor piers comprise four sub-piers and not only two individual piers at the same location. The increase in estimated total scour depth at these three piers as a result of the proposed excavation was estimated to be 0.3 m, 0.6 m and 0.2 m, respectively.

To mitigate the potential impacts of scour (and increases in scour), a design was developed to overcome the issues relating to the estimated scour of the three bridge anchor piers. The design incorporated rock protection works surrounding the piers. The proposed works comprised excavation of the area around the piers, placement of geotextile over the excavated batter, placement of filter rock and rip-rap, and placement of excavated material to reinstate to natural ground level. The design accounted for future undermining by use of a self-launching toe.

The design was based on preventing impacts of estimated additional scour around anchor piers 7, 13 and 19. The following design criteria were adopted:

- The largest increase (impact) on peak velocities – 0.93 m/s (20-year ARI)
- The largest increase in scour – 0.62 m (pier 13)
- The largest peak velocity of 2 m/s (2000-year ARI)

The detailed design of scour protection (following agreement on the basis of design with Roads and Maritime Services) was developed in accordance with well-known manuals. A typical section from the design drawings is presented below in Figure 9.5.

9.2.9 Conclusions

A numerical modelling-based scour assessment was undertaken to estimate scour and inform scour mitigation design. The approach, although containing some limitations outlined below, was considered a useful tool for this application. The numerical modelling results were compared to hand calculations based on empirical equations. The results varied but were generally comparable. The scour estimates are considered conservative, based on the following:

1. The particle sizes used for the sensitivity analysis are representative of silts and clays, which require a lower velocity to induce scour if the effect of cohesion is not considered as is the case with the software used. As some areas of the Nepean Floodplain may contain clays, the cohesive properties these soils contain would help reduce the scour;
2. The numerical model did not take account of the role of vegetation. There is a reasonable turf cover at the site over the floodplain, which could be expected to offer initial resistance to erosion and scour. The role of vegetation in engineering applications is well documented (refer CIRIA/CUR/CETMEF, 2007), as is the role of vegetation (turf) in the resistance of erosion and scour.

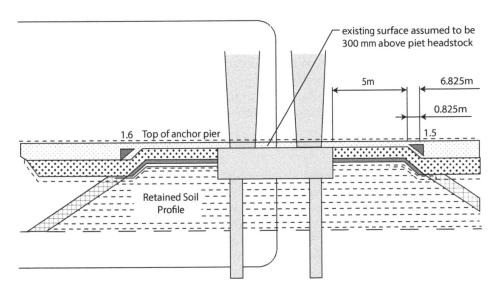

Figure 9.5 **Typical cross-section of scour protection around each anchor pier.**

3. The velocities used in the 1D HEC-RAS assessment were peak velocities. Peak velocities would be experienced for only a proportion of the time, rendering the estimated scour depths conservative.

9.3 Project Waterdunen (Svasek)

9.3.1 Introduction

Scour downstream of a culvert is considered in this case. Immediately downstream of the outflow opening a bed protection has been made, but further downstream scouring is allowed. Section 9.3.2 presents the required bed protection. Section 9.3.3 discusses the flow field downstream which is computed with FINEL2DH (Dam et al., 2016) and the CFD model TUDFlow3D (de Wit, 2015). In Section 9.3.4, scour is estimated with the Breusers formulas adjusted for non-steady flow conditions, taking into account tidal movements. The case concludes in Section 9.3.5 with some additional remarks regarding such items as safety factors and turbulence.

Waterdunen is an intertidal area on the landward side of the dike next to the Western Scheldt estuary in the southwest of the Netherlands. A culvert connects this estuarine nature reserve with the Western Scheldt through the primary sea defence. The tide in Waterdunen is reduced to ±0.75 m +NAP and a smooth variation is guaranteed by automatic gates. The project is an initiative of the province of Zeeland and will become operational in 2020.

On both sides of the culvert, bed protection is required to prevent erosion and instability of the sea defence. However, on the Waterdunen side of the culvert the bed protection has a lower design standard than the primary sea defence as long as the failure of the bed protection does not lead to failure of the sea defence.

The design condition of the bed protection is created by failure of the gates since the gates determine the discharge and flow velocities when the water is flowing into Waterdunen. This means that in case of gate failure the bed protection is allowed to fail and a scour hole to develop. Therefore, this example focuses on the Waterdunen side of the culvert. An impression of the Waterdunen project is depicted in Figure 9.6.

This project was especially challenging because of the high daily discharge through the culvert and large uncertainties in hydraulic loading. This example strives at exemplifying how to deal with these uncertainties in the design process.

9.3.2 Bed protection

The bed protection is designed for spring tide conditions in combination with slight deviations from regular gate operations. The discharge at maximum inflow is approximately 90 m³/s with a water level in Waterdunen of around 0 m +NAP. The bottom of the culvert is located at −2 m +NAP after which a slope of 1:6 is constructed to −2.9 m +NAP ($D/h_0 = 0.3$). The maximum excavation depth is −3.5 m +NAP because of possible explosives at greater depths. This means that the bed protection can have a maximum layer thickness of 0.60 m, assuming a layer thickness of 2.5*d_{n50} for wet construction. The maximum stone grade to be applied is 10–60 kg.

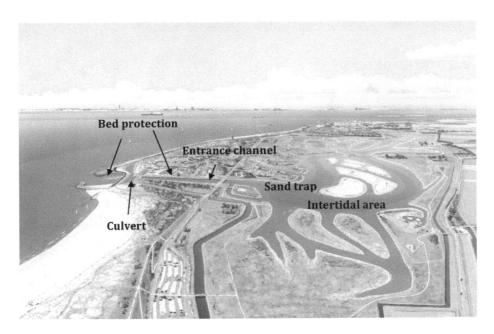

Figure 9.6 Artist impression of project Waterdunen
Source: www.waterdunen.com.

The flow penetrates as a 12 m wide jet into the body of water behind the culvert. The flow velocities behind the culvert are governed by the speed at which the jet mixes over the horizontal plane. In addition, the slope might cause an additional contraction of the flow which would slow down the horizontal mixing.

The flow velocities at spring tide conditions require a stone grade larger than 10–60 kg (determined with the Pilarczyk (2003) design formula). However, a larger grade cannot be constructed because of the maximum allowed layer depth. Therefore, the first 95 m of the bed is protected with penetrated rubble which does not exceed the maximum layer depth and better resists the occurring flow velocities. This penetrated rubble will not fail during the design conditions of the sea defense. However, the bed protection behind this is likely to fail during design conditions and scour can occur (the distance from the culvert $L = 95$ m and the initial water depth $h_0 = 2.9$ m. The corresponding ratio of $L/h_0 = 33$ would lead to $r_0 = 0.12$–0.18.

If failure of the gates occurs the gates will be closed manually within 3.5 h to protect the area from flooding. However, to guarantee the stability of the sea defence within these 3.5 h, the scour may not lead to a flow slide or shear failure with a failure length longer than the bed protection. The allowed limit of the scour depth is 9.5 m (based on a critical failure length of 95 m and assuming a profile after failure of 1:10). To guarantee safety of the sea defence the scour depth within the response time of gate closure must be determined.

Section 9.3.4 briefly describes the determination of the scour depth, but first the derivation of the hydraulic loads is described in more detail.

9.3.3 Hydraulic loads

Two models have been used to estimate the tide-driven flow design conditions:

1. FINEL in 2DH mode to calculate the interaction between outside and inside water levels over the culvert and determine the design situation;
2. TUDFlow3D, a 3D CFD model that resolves turbulent eddies on a small scale to predict the mixing of the jet (plume dispersion) under design conditions. TUD-Flow3D has been developed at Delft University of Technology and extensively validated against scale experiments and field measurements. It has shown good results in resolving plume dispersion (de Wit, 2015).

The 3D model is focused on resolving small-scale turbulent fluctuations (using LES). For this purpose a high-resolution grid with $dx = 0.25$ m, $dy = 0.25$ m and $dz = 0.18$ m is used. The only disadvantage is that the model works with a rigid-lid assumption which means that the water level is constant throughout the domain in the equations for mass conservation. This assumption can lead to small deviations in velocities (<10% beyond the penetrated rubble bed protection).

The 3D model shows a longer jet, sticking to the eastern side slope, and higher velocities compared to the 2D model. After 100 m the jet starts mixing over the width of the channel. Figure 9.7 shows the depth-average velocity according to FINEL (left panel) next to the time- and depth-averaged velocity (middle panel) according to the 3D model and the depth-averaged maximum velocity (right panel) according to the 3D model.

The 3D model resolves turbulent vortices at a small scale. This means that local velocities vary in time and space. Figure 9.8 shows the instantaneous depth-averaged velocity (instantaneous because the velocity is only averaged over the depth, not over time). The irregular shape of the jet shows that turbulent vortices occur in the mixing layer of the jet and within the main flow. The relative turbulence intensity, r_0, can be directly derived from the calculation results to provide guidance on estimating this value to which the results are sensitive.

These results show that the 2D model by itself underestimates the intensity of the flow velocities because the outflowing plume is mixed over the entire width of the canal almost instantly. For this reason, the results from the 3D model have been used for the design. The results from the 2D model have been used for sensitivity and scour scenarios after correction for the underestimated mixing length.

9.3.4 Scour depth

Scour is expected to occur at the end of the penetrated bed protection only when the gates fail. At the end of the bed protection, the channel is wider and a deep sand trap has been constructed that reduces flow velocities to below the critical velocity. So no scour is expected.

Immediately after failure of the gates flow velocities will be low because the hydraulic head is still relatively low. Yet, velocities will increase until high water at the Western Scheldt has been reached or the gates have been closed. The speed of scour development depends strongly on this variation in time of the flow velocity. The characteristic time, defined as the time in which the scour depth reaches a depth equal to the initial water depth, can be calculated with Equation (6.15):

$$t_{1,u} = \frac{Kh_0(0)\Delta^{1.7}}{\dfrac{1}{T}\displaystyle\int_{t1}^{t2} \dfrac{\left(\alpha U_0(t)-U_c\right)^{4.3}}{h_0(t)}\,dt} \tag{6.15}$$

where
$$\alpha = 1.5+4.4\,r_0 f_c\,(-)$$
$$f_c = C/C_0\,(-)$$

C_0 is assumed to be $40\,\mathrm{m}^{1/2}/\mathrm{s}$ and C is the Chézy coefficient (here about $65\,\mathrm{m}^{1/2}/\mathrm{s}$). The relative turbulence intensity (r_0) in non-uniform flow with high turbulence varies between 0.15 and 0.20 (see Table 4.3). Considering the 3D results, a conservative value of 0.2 is assumed. This is conservative because $r_0 = 0.12$–0.17 for $L/h_0 = 33$ in Figure 6.7.

The critical flow velocity can be determined with:

$$U_c = 2.5\,\sqrt{\psi_c \Delta g d_{50}}\;\ln\!\left(12h/k_s\right)$$

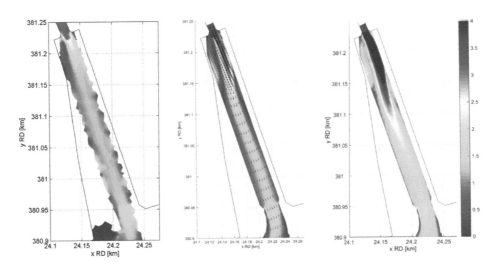

Figure 9.7 Design depth-averaged velocity according to FINEL (left), and time- and depth-averaged velocity (middle) and depth averaged maximum velocity (right) according to TUDFlow3D. The colour scale is equal in all cases.

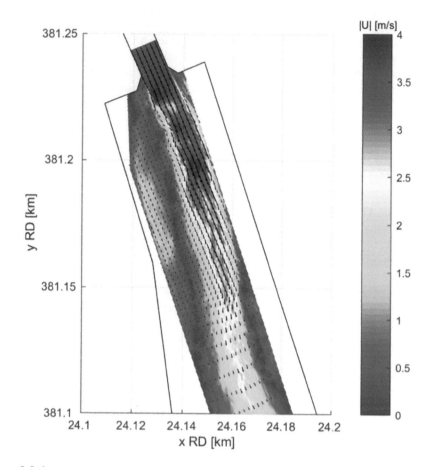

Figure 9.8 Instantaneous depth-averaged flow velocity for the first 150 m after outflow of the culvert.

The grain size is uncertain but assumed as $d_{50} = 187\,\mu$m based on the limited data
available. Assuming a kinematic viscosity $v = 1*10^{-6}$ m^2/s and $\rho_s = 2650$ kg/m^3 and
$\rho = 1025$ kg/m^3, it follows that $\Delta = 1.59$ and $D_* = 4.7$.

The critical mobility parameter is determined by (see Table 4.4):

$$\psi_c = 0.14 D_*^{-0,64}$$

For a water depth h of 2.9 m and a Nikuradse roughness (k_s) of $3*d_{90} = 900\,\mu$m, U_c
follows as 0.31 m/s.

With the characteristic time, the development of the scour hole over time can be
determined:

$$\frac{y_m(t)}{h_0(0)} = \left(\frac{t}{t_{1,u}}\right)^\gamma \tag{6.14}$$

in which γ is assumed at a conservative value of 0.5 compared to the values in Table 6.1 for two-dimensional (2D) flow. Figure 9.9 presents the velocity just upstream of the scour hole and the resulting scour depth growth, calculated with Equation (6.14). The scour depth is expected to reach 8 m in 3.5 h with a characteristic time scale of 0.5 h. Note that this is a conservative estimate since Equation (6.14) is only applicable as long as $t < t_{1,u}$. After the first half hour, the scour depth growth is likely to decrease.

Using Equation (6.14), with $r_0 = 0.12$ instead of $r_0 = 0.2$ would have led to a maximum scour depth estimate of only 5 m in 3.5 h. More importantly, if the velocity of the 2D calculation were used to predict the maximum scour depth, the estimate would have been 3 m instead of 8 m based on the 3D prediction. These results show how sensitive the estimated scour depth is to the flow velocity, highlighting the importance of a thorough investigation of the expected hydraulic loads.

9.3.5 Additional remarks

9.3.5.1 Gate control

The hydraulic loading on the bed protection depends on the gate control. Deviations in gate control can increase hydraulic loading significantly at inflow since the culvert capacity is multiple times higher than that required at inflow (but required at outflow).

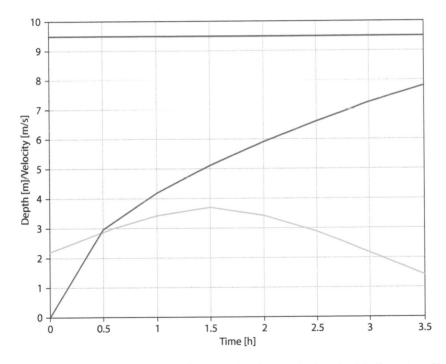

Figure 9.9 Conservative estimate of scour depth as calculated with Equation (6.14) (blue line), development in hours after gate failure and flow velocity just upstream of scour hole (green line). Within 3.5 hours the gate will be closed so no exceedance of the critical scour depth occurs. The red line indicates the allowed limit scour depth before the sea defence fails.

9.3.5.2 Safety factors

Design safety is accounted for via both loading and strength. The strength is increased by 10% by decreasing the allowable scour depth limit. The hydraulic loads are also increased by 10% to account for model and measurement uncertainties.

9.3.5.3 Sensitivity calculations

Sensitivity calculations with respect to the scour depth have been made to assess the sensitivity of the design to both hydraulic loading, design formulas and parameter settings.

9.3.5.4 Turbulence

An important factor is the turbulence intensity. The 3D CFD model TUDFlow3D has been used to support the choice of the turbulence factor. This plume model resolves turbulent eddies on a small scale and therefore the depth-averaged relative turbulence intensity $r_0 = \sigma_u/U$ can be directly calculated from the model results instead of estimated based on Figure 6.7.

9.4 Full-scale erosion test propeller jet (Deme)

9.4.1 Introduction

Jets induced by main propellers of ships may result in scouring. The present case addresses this. The scour due to the two main propellers can be predicted with several formulas: the erosion rate formula of Osman and Thorne (1988) and Thorne (1993) mentioned in Section 3.5.4, the erosion rate formula of Briaud (2008) and the Römisch and Schmidt (2012) formula mentioned in PIANC report 180 (PIANC, 2015). The latter report is recommended in this scour manual (see Section 5.7). In essence, the Römisch and Schmidt formula was derived for situations with a propeller jet attacking a horizontal bed and therefore is not applicable for a propeller jet hitting a slope.

The case compares the predicted scour depths with the various formulas and with full-scale in-situ tests. These tests were carried out because the results of the predicted scour depths were unrealistic. The case shows that the scour formula of Römisch and Schmidt, and also the erosion rate formulas of Thorne and Briaud, do not result in reliable results. Only by adjusting the coefficients in the formulas considerably the scour can be predicted correctly.

First, the computed flow field is described. Then, the scour prediction methods are outlined, and finally the results of the computed and measured scour depths are presented.

9.4.2 Objective of the full-scale erosion tests and estimated flow field

As part of a life-cycle cost analysis for the structures of a harbour basin, a full-scale propeller-wash erosion test was undertaken on a dredged slope consisting of heterogeneous subsoil conditions using a test vessel (Figure 9.10). The objective of the test was to estimate the erosion rate of the in-situ material in order to predict the final profile of the dredged slope at the end of the design life following exposure to propeller wash (see Karelle et al., 2016). The test was motivated by the large predicted scour depths using the formulas following PIANC report 180 (PIANC, 2015). It should be noted that the velocity profile was determined based on a calculated flow field determined by the Dutch approach (PIANC, 2015) (Figure 9.11).

9.4.3 Scour prediction methods

Briaud (2008) provides erosion rate curves, R (in mm/hr), for different soil materials which can generally be written as:

$$R(U) = aU^\beta \tag{4.26}$$

with α and β coefficients depending on the in-situ soil and U is the flow velocity in mm/s. Based on the test results, a best-fitting curve for the material was found with coefficient $\alpha = 17$ and exponent $\beta = 2.9$ (Karelle et al., 2016). Note that these values differ from the values given by Briaud: $\alpha = 1.2$ (clay) to 11.8 (sand) and $\beta = 2.9$ (clay) to 5.6 (sand).

We compare the results of the in-situ test with the formulae of Römisch and Schmidt (PIANC, 2015) for non-cohesive soils and the method for cohesive soils according to Osman and Thorne (1988) and Thorne (1993) (see Section 4.4.5). The formulas of Thorne are:

$$\frac{dz}{dt} = \frac{R}{\rho_b \cdot g} \cdot \tau_c \left(\frac{\tau_b}{\tau_c} - 1 \right) \text{ and } R = 0.364 \cdot \tau_c \cdot e^{-1.3 \cdot \tau_c} \tag{4.24}$$

$$\tau_b = \frac{1}{2} \cdot c_f \cdot \rho \cdot u_b^2 \tag{4.17}$$

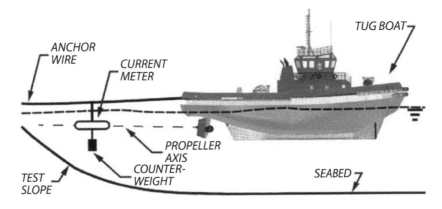

Figure 9.10 **Test setup.**

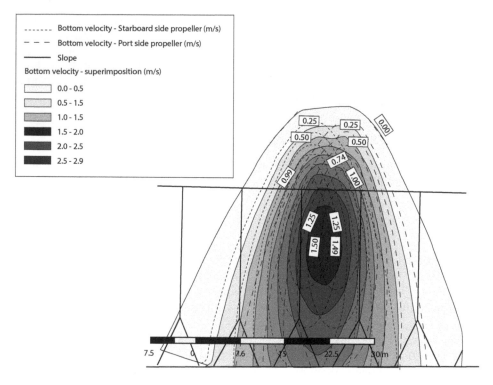

Figure 9.11 Theoretical flow field at the slope based on Dutch approach (PIANC, 2015).

where

R = erosion parameter (kg/(m·s³)),

t = time (s),

ρ_b = density of the bed material (kg/m³),

c_f = resistance coefficient (–), c_f = range 0.005–0.020,

u_b = flow velocity just above the bed (m/s),

ρ = density of water (kg/m³).

The value of τ_c can be computed with (Winterwerp et al., 2012):

$$\tau_c = 0.7 PI^{0.2} \quad \text{for} \quad PI < 0.07 \tag{4.20}$$

or

$$\tau_c = 0.163 PI^{0.84} \quad \text{for} \quad PI < 0.07 \tag{4.21}$$

where PI = plasticity index (–).

The formulae for main propellers of Römisch and Schmidt considers long-term exposure and are shown below:

$$\frac{y_{m,e}}{d_{85}} = \frac{h_p}{d_{85}} C_{ad} C_{mr} \left[a_a \frac{B}{B_c} - 1 \right]$$ (9.3)

and

$$B = \frac{U_{bed}}{\sqrt{d_{85} g \Delta}}$$ (9.4)

where
d_{85} = characteristic bed diameter (m)
h_p = distance between propeller axis and bed (m)
C_{ad} = coefficient (–)
C_{mr} = coefficient (–)
a_α = coefficient (–)
B = load factor (–)
B_c = stability factor (–)
U_{bed} = jet velocity at the bed (m/s)

We recall that the Römisch and Schmidt equation was derived for a horizontal bed, not for a slope.

9.4.4 Results

We used an exposure time of about 500 hours in the formulae of Briaud in order to simulate the long-term results of Römisch and Schmidt. The real duration of the in-situ tests were 22 hours. Furthermore, we adjusted the values of C_{ad} and $C_{m,r}$ in the Römisch and Schmidt formula to obtain results comparable with the in-situ tests. The following parameters were used for the comparison:

Römisch and Schmidt	Thorne
$B_c = 1.2$	$\rho_b = 1700$ kg/m^3
$C_{ad} = 1.0$	$R = 0.0005$
$C_{m,r} = 0.3$	$c_f = 0.02$
$a_\alpha = 0.65$	$PI = 4$
$h_p = 5.2$ m	$\tau_c = 0.52$ N/m^2

The following parameters were taken into account for sand and clay (the original values presented by Briaud) and for the heterogeneous sand/clay mixture in the DEME in-situ test results.

Soil type	α	β
Clay (original values Briaud)	1.2	2.9
Sand (original values Briaud)	11.8	5.6
Sand/clay mixture (DEME in-situ test)	17	2.91

Figure 9.12 shows the flow velocity near the bed versus the computed erosion rate, according to the formulas of Briaud and Thorne, and the flow velocity near the bed versus the measured erosion rates. The maximum measure erosion rate occurs immediately after the start of the tests. The erosion rate will decrease to zero at the equilibrium scour depth.

Figure 9.13 shows the computed scour depths according to different prediction formulas and the in-situ measurements. The presented in-situ scour depths are computed with the measured erosion rates times the 500 hours. Thus, these are not real measured scour depths but virtual ones to enable a comparison with the prediction formulas. The real measured maximum scour depth is between 1 and 3 m (Karelle et al., 2016).

Furthermore, we have to mention that only computed virtual scour depths up to 4.5 m are shown being already deeper than the really measured scour depth.

We conclude that the Römisch and Schmidt formula should be considered as a long-term scour development formula which gives very conservative results. The recommended value is $C_{m,\,r} = 0.3$–0.44 for manoeuvring situations and 1.0 for stationary situations. As the vessel was kept at the same location, the test case would fall into the stationary situation. Using the advised parameters of the PIANC report 180, a factor 10 higher scour holes are calculated than measured with the full-scale test. Thus, the Römisch and Schmidt formula is not applicable for predicting the erosion on a slope due to a propeller jet.

The formula of Thorne is close to the results calculated with Briaud for clayey materials, which is consistent with the fact that this can be used for cohesive materials. Comparing their predictions with the results from the test, we conclude that the formulas of Thorne and Briaud are too optimistic. This stands to reason because the in-situ test was carried out with heterogeneous materials.

9.5 Scour due to ship thrusters in the Rotterdam port area (Port of Rotterdam)

9.5.1 Introduction

This case describes the scour at locations in the Rotterdam port area, in particular scour at quay walls. Section 9.5.2 presents results of full-scale tests with inland vessels. In Section 9.5.3, the observed scour at a quay wall for sea-going container vessels is compared with results of prediction formulas. Both cases will show that the available

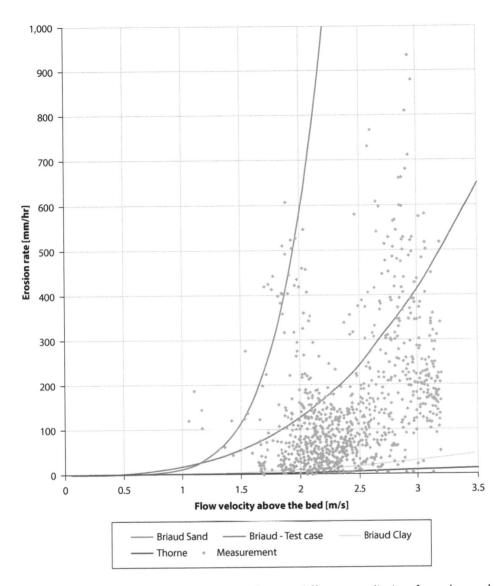

Figure 9.12 Computed erosion rate according to different prediction formulas and in-situ measured erosion rate (mm/h).

formulas are not adequate for predicting scour, but it should be kept in mind that full-scale measurements themselves might be inaccurate too, for example with respect to the test conditions. Nevertheless, both cases show that the formulas recommended in PIANC report 180 (PIANC, 2015) underestimate the real scour depth, whereas the Hoffmans formulas overestimate the scour depth.

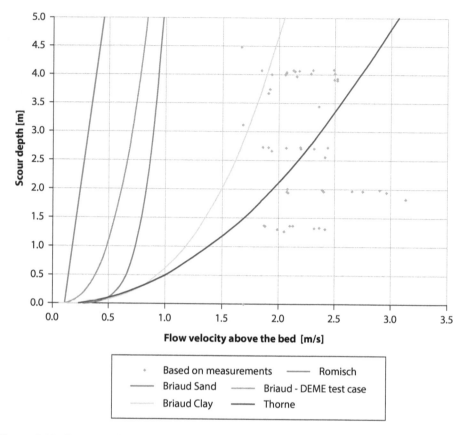

Figure 9.13 Computed scour depth according to different prediction formulas and computed scour depth based on the in-situ measured erosion rate (500 hours).

9.5.2 Full-scale test with inland vessels at the Parkkade

9.5.2.1 Scope

Prototype tests at the Parkkade quay along the river Nieuwe Maas in Rotterdam were carried out to assess the scour caused during departures of two fully loaded inland tankers: barge N with dimensions 110 × 13.5 × 4.1 m and barge J with dimensions 135 × 20 × 3.85 m). The scour caused by bow thrusters and the main propeller, directed perpendicularly towards the quay, was measured.

Four test series were performed. Each series consisted of 1–6 departures of one vessel from the same starting position along the quay wall. The departures were performed in such a way that the movement of the vessel was perpendicularly away from the quay wall, without rotation and forward or backward movement of the vessel. Each departure manoeuvre finished when the vessel had moved over 30 m away from the quay wall. If one bow thruster was used, this manoeuvre took about 3 minutes and with two thrusters about 1.5 minutes. This type of departure manoeuvre will normally be applied only if other vessels are moored directly in front of and behind the departing vessel. This manoeuvre causes the largest erosion in front of the quay because during the whole manoeuvre, the maximum hydraulic load on the bottom in front of the quay remains at the same location.

Both barges were equipped with two bow thrusters. Barge N was equipped with two 4K channel thrusters, whereas barge J was equipped with one 4K channel thruster and one Compact Jet (CJ) thruster (Figure 9.14). The principle of the 4K type is that water is vertically sucked up from under the vessel through a horizontally lying propeller. The water jet is then deflected 90° and guided through one of four channels. The sideward channels are horizontal. The forward and backward channels make an angle of about 20° with the keel of the vessel. The principle of the CJ type is similar. However, the water is not guided through channels, but the outflow is oblique downwards directly next to the inflow of the thruster. The outflow makes an angle of 25°–30° with respect to the keel of the vessel. The Compact Jet is horizontally rotatable over 360°.

Barge N was equipped with one fixed main propeller and a rudder, whereas barge J was equipped with two azimuthal (360° rotatable) main propellers.

9.5.2.2 Observed scour depth versus predictions with Breusers formulas

Table 9.5 shows the measured scour depth for single departures and for a whole series of departures. The scour caused by one thruster during one single departure has a maximum depth of 0.5 to 0.6 m. The total maximum scour depth caused by series N1, consisting of six departures with one 4K channel thruster using 320 kW, is about 1.8 m.

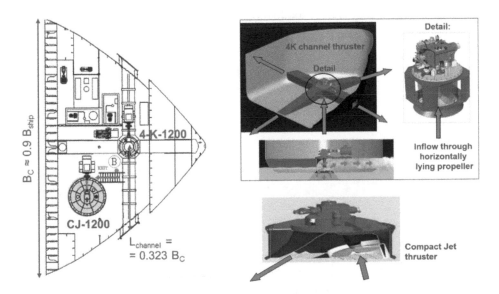

Figure 9.14 **Bow of barge J (left), 4K channel thruster (upper right), and Compact Jet thruster (lower right).**

Table 9.5 Measured scour depths per departure and per series

| Series: | Scour by thrusters | Scour by sideward rotated main propeller | | | | | |
| | Barge N | Barge J | | | | | |
	N1	J1	J2	J3	J1-M	J2-M	J3-M
Thruster type:	1×4K	1×4K	1×CJ	4K + CJ			
Dep.nr.	Scour depth per departure and summed (cm)						
1	60	40	40–50	45	48	30–35	70
2	20	15	45–50		15	15–10	
3	?	55			15		
4	?	0			5		
5+6	?						
SUM	180	110	95	45	83	45	70

Figure 9.15 shows the scour depth as a function of the accumulated duration of successive departure manoeuvres for different series, with two dotted linear regression lines for the series N1 and J1. The duration of one departure manoeuvre is about 3 minutes in which the vessel moves about 30 m away from the quay wall. However, the duration of each departure has been corrected to 1.25 minutes in order to make a correct comparison with results from formulas of Breusers, which are also shown in Figure 9.15. The calculations of the time development of scour with formulas of Breusers were performed for the maximum hydraulic load occurring at the beginning of the manoeuvre (when there is no distance between vessel and quay wall). Actually during the manoeuvres the hydraulic load ($U_{b, max}$) on the bottom in front of the quay wall

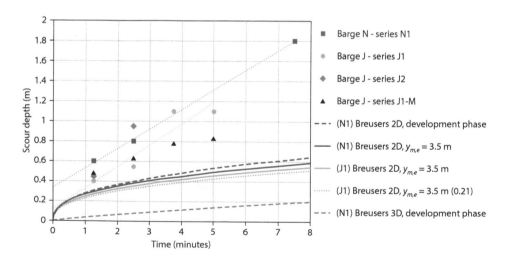

Figure 9.15 Development of scour in time.

gradually decreases. This decrease was implemented in the calculations by reducing the duration of the departure manoeuvres from 3 to 1.25 minutes. This time correction had been derived assuming that the rate of erosion is proportional to $U_{b,\,max}^{2.25}$, just as y_m in Equation (9.5).

The measured scour depths in Figure 9.15 show for most series no asymptotic approach to an equilibrium scour depth, except for series J1-M which tends to an equilibrium scour depth of 1.0 m. Extrapolation of the dotted trend lines for the series N1 and J1 would result in a significantly larger equilibrium scour depth than the measured maximum scour depths of 1.8 m and 1.1 m. However, the number of well surveyed departures is insufficient to draw reliable conclusions about the equilibrium scour depths. Unfortunately, the bottom surveys after departures 3, 4 and 5 of series N1 failed due to reflection of survey beams close to the quay wall.

In addition, Figure 9.15 shows the time development of scour according to Equation (3.6) of Breusers and according to Equation (3.7) for the development phase of scour. The time scale t_1 was calculated with Equation (3.9). The starting points for the calculations with these equations were: $\gamma = 0.4$ for 2D flow, $\gamma = 0.8$ for 3D flow, $\lambda = 0.3 \cdot (x_{pq} + h_{pb})$ (see Table 9.6 for values), U_0 in Equation (4.9) = $U_{b,max}$, according to the Dutch method (PIANC, 2015) for a jet against a quay wall, $r_0 = 0.3$, $U_c = 0.3$ m/s, $\Delta = 1.65$, $y_{m,e} = 3.5$ m (estimated based on measured data, see Figure 9.14). This results in $t_1 = 2.54$ hours for barge N (series N1) and $t_1 = 7.06$ hours for barge J (series J1).

Figure 9.15 shows that the formulas of Breusers with the aforementioned starting points underestimate the measured scour depths due to jets redirected by a quay wall to the bed, but it should be kept in mind that the Breusers formulas were not derived for a redirected jet. For the development phase, Breusers 2D gives a much better estimate than Breusers 3D. However, for $t > t_1$ the scour according to Breusers 3D is larger than according to Breusers 2D (not shown in figure).

9.5.2.3 Observed versus predicted scour for thrusters with PIANC formulas

Figure 9.16 shows measured scour holes caused by the thruster jet of series N1. Figure 9.17 shows the difference between bottom surveys for series J1 and J2 before and after all vessel departures. The figures show that the surface area of the scour hole caused by the Compact Jet (which is directed obliquely towards the bottom) is significantly larger than the surface area of the scour hole caused by the 4K channel thruster.

The measured maximum scour depths caused by the 4K channel thrusters are compared with results of the formulas of Römisch and Schmidt for scour caused by a thruster jet perpendicularly impinging on a quay wall (Römisch & Schmidt, 2012; PIANC, 2015):

$$\frac{y_{m,e}}{d_{85}} = 4.6\, C_m \left(\frac{B}{B_c} \right)^{2.25} \quad for \quad \frac{B}{B_c} > 1.4 \tag{9.5}$$

and

$$B = \frac{U_{b,max}}{\sqrt{d_{85} g \Delta}} \tag{9.6}$$

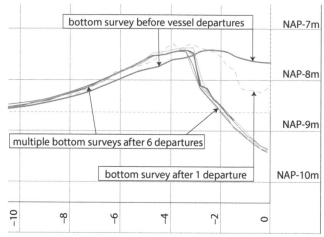

distance from upper part of combined quay wall (m)

Figure 9.16 Scour after 6 departures with 4K channel thruster with 320 kW (series N1).

Note: scour contour lines every 0.1 m.

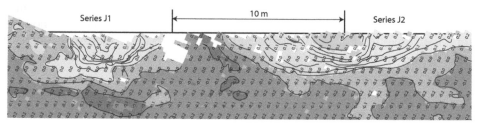

Legend: contour lines every 0.1 m

Figure 9.17 Scour by 4K thruster (series J1) and CJ thruster (series J2).

with:

B_c = stability coefficient,

B_c = 1.2 for bow thrusters (–),

C_m = coefficient for duration of the jet load (–),

C_m = 1.0 for stationary load; C_m = 0.3 for temporary loads during manoeuvres.

The main input parameter in these formulas is the maximum flow velocity near the bottom in front of the quay ($U_{b, max}$). The value of $U_{b, max}$ is calculated by two different methods: the German method and the Dutch method in PIANC Report 180 (PIANC, 2015).

Table 9.6 shows the input parameters and the results of calculations of $U_{b, max}$ with the German and the Dutch methods and calculated values of the maximum scour depth ($y_{m, calc}$) based on the calculated values of $U_{b, max}$. Scour depths are calculated

Table 9.6 Calculation of scour depth at Parkkade (d_{85} = 320 µm)

Parameter	Unity	Barge N (NI)	Barge J (JI)
Number of departures	(–)	6	4
Measured scour depth ($y_{m,meas}$)	**(m)**	**1.80**	**1.10**
Non eroded bottom level (NAP)	(m)	–7.70	–8.00
Draught of vessel	(m)	4.10	3.85
4-channel-bow thruster			
Power applied	(kW)	320	405
Equivalent circular channel diameter D_0	(m)	0.91	0.91
distance thruster-quay(x_{pq})	(m)	3.25	5.19
distance thruster-bottom (h_{pb})	(m)	3.71	4.36
Scour with flow velocity according to German method			
Outflow velocity U_0	(m/s)	8.37	9.06
$U_{b,max}$	(m/s)	2.67	1.00
Calculated scour ($y_{m,calc}$) (C_m=0.3)	**(m)**	**1.00**	**0.11**
$y_{m,meas}/y_{m,calc}$	(–)	1.80	10.1
Scour with flow velocity according to Dutch method			
outflow velocity U_0	(m/s)	7.67	8.29
$U_{b,max}$	(m/s)	1.88	1.73
Calculated scour ($y_{m,calc}$) (C_m=0.3)	**(m)**	**0.45**	**0.38**
$y_{m,meas}/y_{m,calc}$	(–)	4.0	2.9
Scour with Equation (5.20)			
$y_{m,calc}$	**(m)**	**3.65**	**3.64**
$y_{m,meas}/y_{m,calc}$ (6.27)	**(–)**	**0.49**	**0.30**

with Equation (9.3) using $C_m = 0.3$, and this value is recommended for manoeuvring vessels. The calculated scour depth is about 0.4 m applying the Dutch method and 0.1 to 1.0 m applying the German method.

The end of Table 9.6 also shows calculated scour depths according to Equation (5.22) for 3D culverts (Hoffmans, 2012), applying $Q = 1/4 \cdot D_0^2 \cdot U_0$, $U_1 = U_{b, max}$ according to the Dutch method for a jet against a quay wall, $U_2 = 0$ m/s, $d_{90} = 330$ µm and 10°C water temperature for the calculation of D_{90*}, resulting in $c_{3H} = 3.7$. This calculation with Equation (5.22) results in an equilibrium scour depth $y_{m,e} = 3.6$ m. Hoffmans (2012) also mentions an alternative value for c_{3H}, namely $c_{3H} = 1.6$, based on the research of Ruff et al. (1982). This value of c_{3H} results in $y_{m,e} = 1.6$ m.

Table 9.6 shows also ratios between the measured and the calculated scour depths.

More information on the Parkkade tests can be found in Roubos et al. (2014) and Blokland (2013).

9.5.2.4 Conclusions

The maximum measured scour depth (1.8 m) after six departure manoeuvres is a factor 4.0 larger than the scour depth which follows from Equation (9.5) (Römisch & Schmidt, 2012) with $C_m = 0.3$ for manoeuvring vessels, when $U_{b,max}$ is calculated with the Dutch method.

When $U_{b,max}$ is calculated with the German method, the ratio between measured and calculated erosion is a factor 1.8 after six manoeuvres with barge N but a factor 10 after four manoeuvres with barge *J*. So the results based on the German method for $U_{b,max}$ are not consistent.

Equation (5.22) (Hoffmans, 2012) with $c_{3H} = 3.7$ results in a maximum equilibrium scour depth which is a factor 2 larger than the measured maximum scour depth after six manoeuvres.

Almost certainly, the measured maximum scour depth is (significantly) smaller than the equilibrium scour depth for more repetitions of the same manoeuvre at the same location. However, it should be noted that operational use of a quay wall generally implies varying manoeuvres at varying locations. A reliable conclusion about the equilibrium scour depth for operational use by inland vessels cannot be drawn.

The 2D and 3D formulas of Breusers (Equations 3.6–3.8) result in an underestimation of the measured scour depths. For the development phase of scour, Breusers 2D gives a much better estimate than Breusers 3D.

9.5.3 Scour due to operational use of Maasvlakte quay wall for large seagoing container vessels

This section discusses an example of scour due to operational use of a quay wall for large seagoing container vessels. The quay wall, which is located in the Maasvlakte area of the port of Rotterdam, was delivered in 2010 with a nautically guaranteed bottom level of 16.65 m +NAP. Accounting for future developments, the quay wall was designed for a bottom level of −22.00 m +NAP. No bottom protection was installed because the delivered bottom level was much higher than the design level.

9.5.3.1 Observed scour

Figure 9.18 shows the bottom profile which was measured from 2013 to 2017 in a section parallel to the quay wall at 2 m in front of the quay wall. The largest container vessels have more or less fixed mooring positions. As a result, the scour by thruster and propeller jets always occurs in the same areas. These scour areas (A, B and C) can be recognised in the bottom profile along the quay wall.

The characteristics of the largest mooring container vessels are: length 400 m, width 58.8 m and design draught 14.5 m. The characteristics of its bow thrusters are: 2 thrusters, 2500 kW per thruster and diameter $D_0 = 2.7$ m. In moored position, the distance ($L = x_{pk}$) between the outflow of the thruster and the quay wall is 24.4 m. The keel clearance, averaged over all manoeuvres of this class of largest container vessels, is 2.0 m. The minimum keel clearance is about 0.5 m. For these two clearances, the height of the propeller axis above the non-eroded bottom (h_{pb}) is 5.35 m and 3.85 m, respectively. It is not evident which value of the keel clearance should be chosen for the calculation of the scour depth: the minimum clearance with largest hydraulic load but a short duration, or the mean clearance with less hydraulic load but longer duration. The best would be to calculate the scour for each separate manoeuvre with actual keel clearance and actual thruster use. However, this is a very elaborate method, which is moreover not feasible because there is no formula available for the rate of scour during separate manoeuvres (depending on the already existing scour hole).

Figure 9.18 shows that the maximum scour depth measured until 2017 is about 2.75 m (at location A). The development in time of the maximum scour depth at A, B

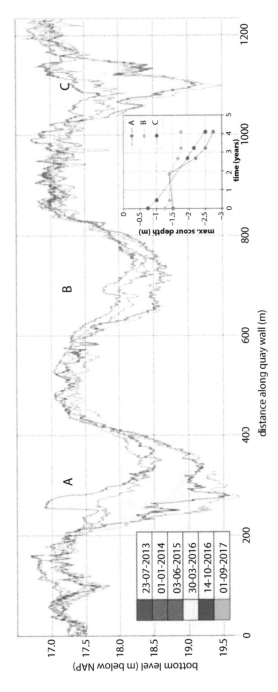

Figure 9.18 Measured bottom level along section at 2 m in front of quay wall.

and C is shown in the small graph added to Figure 9.18. Extrapolation of the graph for location A would result in a maximum equilibrium scour depth of 3.0 m. The graph for location B shows no approach to an equilibrium. A reliable conclusion about the maximum equilibrium depth cannot be drawn. Therefore, in the following, calculated scour depths will be compared with a measured maximum scour depth of 2.75 m ($y_{m,meas} = 2.75$ m).

9.5.3.2 Computed scour

Table 9.7 shows calculated values of the maximum scour depth ($y_{m,calc}$) and ratios between the measured and the calculated scour depths. The scour depths are calculated using Equation (9.5) with $C_m = 0.3$, resulting in a scour depth of 1.3–1.4 m according to the Dutch method and in 0.2–0.4 m with the German method.

For $U_{b,max}$ calculated with the Dutch method, the scour depth was also calculated with Equation (5.22) (Hoffmans, 2012) with $U_0 = 8$ m/s, applying for the two bow thrusters $Q = 2 \cdot (1/4) \cdot D_0^2 \cdot U_0$, $U_1 = U_{b,max} = \sqrt{2} \cdot U_{b,max,single}$, $U_2 = 0$ m/s, $d_{90} = 270$ μm, and 10°C water temperature for the calculation of $D90^*$, resulting in $c_{3H} = 3.9$. This calculation results in an equilibrium scour depth $y_{m,e} = 11$ m. Hoffmans (2012) also mentions an alternative value for c_{3H}, namely $c_{3H} = 1.6$, based on research of Ruff et al. (1982). This value of c_{3H} results in $y_{m,e} = 4.5$ m.

9.5.3.3 Conclusions

The maximum measured scour depth after several years of operational use is 2.75 m. Probably, the equilibrium scour depth is somewhat larger considering the scour development in time at the different locations.

Table 9.7 Calculation of scour depth at Maasvlakte quay wall ($d_{85} = 250$ μm)

Keel clearance	0.5 m	2.0 m
Scour with Equation (9.5) and $U_{b,max}$ according to the German method		
$U_{b,max}$ according to the German method	1.74 m/s	1.20 m/s
$y_{m,calc}$ Equation (9.5) ($C_m = 0.3$)	0.41 m	0.18 m
$y_{m,meas}/y_{m,calc}$	6.7	15.5
Scour with Equation (9.5) and $U_{b,max}$, according to the Dutch method		
$U_{b,max}$, according to Dutch method	3.02 m/s	2.87 m/s
$y_{m,calc}$ Equation (9.5) ($C_m = 0.3$)	1.42 m	1.26 m
$y_{m,meas}/y_{m,calc}$	1.9	2.2
Scour with Equation (5.22) and $U_{b,max}$, according to the Dutch method		
$U_{b,max}$, according to the Dutch method	2.72 m/s	2.58 m/s
$y_{m,calc}$ Equation (5.22) ($c_{3H} = 3.9$)	11.1 m	10.9 m
$y_{m,meas}/y_{m,calc}$	0.25	0.25

The measured maximum scour depth is larger than that computed using Equation (9.4) (Römisch & Schmidt, 2012) with $C_m = 0.3$ for manoeuvring vessels. The Hoffmans (2012) method with $c_{3H} = 3.9$ results in a maximum equilibrium scour depth which is also larger than the measured maximum scour depth.

It can be concluded that the available prediction formulas for scour due to thruster jets seem not very reliable. Note that in case of the very large container vessels, the outflow of the thrusters is so far away from the quay that the thruster jets hit the bottom before they collide with the quay wall. This implies that in the collision with the quay wall, the thruster jet mainly will be reflected sideward along the wall and less or not backward over the bottom. It also implies that the formulas for $U_{b, max}$ and scour depth are possibly less reliable because the formulas are based on measurements in the situation of backward reflection.

9.6 Crossing of high voltage power line (Witteveen & Bos)

9.6.1 Introduction

Scouring around multiple piers below a footing are addressed in this case. After presentation of scope and flow conditions hereafter, the methods to compute scour for a single pier and multiple piers are presented in Sections 9.6.2 and 9.6.3, respectively. For a single pier the HEC-18/CSU relation, Equation (8.9) is applied. For the multiple-pier configuration, formulas are applied to take into account the oblique flow direction relative to the pier configuration. In Section 9.6.4, the results are presented and discussed.

In order to strengthen the Belgian electricity network, high-voltage power lines need to cross the Scheldt. To this end, two large power pylons are placed in the river bed, each consisting of four pile groups which consist of eight piles holding a footer, Figure 9.19. The expected scour is evaluated per set of two pile groups that are aligned in the streamwise direction. The governing storm surge condition has been determined using the SCALWEST model, a WAQUA software-based model specifically made to calculate the water and salt movement in the Scheldt estuary. These calculations were validated with measured bed level heights, water levels and flow velocities, leading to the conditions specified in Table 9.8.

The diameter b of the circular piles is 1.07 m, the distance between them $S1$ is 2.70 m and the distance between the groups of piles $S2$ is 26.40 m.

9.6.2 Scour for a single pier

First, the scour depth for a single pile has been calculated with the HEC-18/CSU relation (Equation 8.9). For the circularly shaped piers, a shape factor K_s of 1.0 was applied. Circular piers are neutral with respect to the flow direction, hence K_ω is also 1.0. HEC-18 also takes into account a factor for the bed condition, denoted as K_3. This factor is set to 1.1 for small dunes or plane-bed conditions (cf. Table 7.3 of HEC-18). This factor is not mentioned in Section 8.5 which, on the contrary, does discuss the factor K_g to account for the influence of the gradation of the bed. This factor is a function of the geometric standard deviation σ_g with a maximum of 1.0, as shown in Figure 8.15.

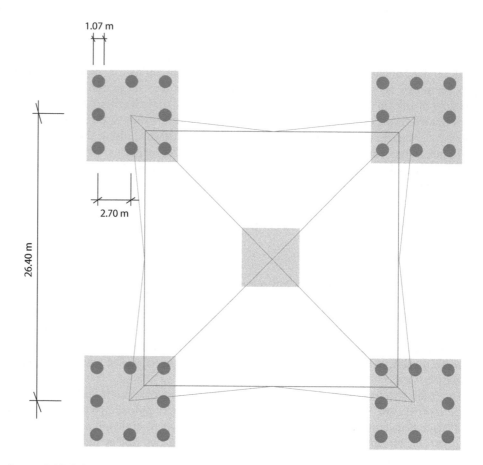

Figure 9.19 Schematic top view of one pylon.

Table 9.8 Water depth and flow velocity per location

$Location_{pile}$ group	Maximum water depth h_0 (m)	Maximum flow velocity U (m/s)
$1_{1/2}$	1.50	0.5
$1_{3/4}$	3.62	0.5
$2_{1/2}$	4.52	1.0
$2_{3/4}$	2.60	1.0

Note: 1_1/2 means pylon 1 with pile groups 1 and 2 upstream

It is important to notice that apparently the K_i in different formulas does not always refer to the same product of factors.

With $K_s = 1$, $K_\omega = 1$ and $K_3 = 1.1$, $K_i = 1.1$. Along with the governing flow conditions shown in Table 9.8, the upper limits for the scour hole depth can be determined. The results are shown in the second and third column of Table 9.9. Note that for pile group 2, the ratio $y_{m,e}$ over b is greater than the value of 1.4 used as a first estimate for live-bed scour (see Equation 8.5), whereas for pile group 1, it is smaller, mainly due to the lower flow velocities.

9.6.3 Scour for multiple piers

The distance between the piles within a pile group is smaller than three times the diameter b which means that the influence between the piles is assessed based on Section 8.5.5. While Section 8.5.5 only offers relations for a set of two piles, HEC-18 uses a formula to determine the so-called equivalent or effective width of a pile group a_{pg}^*, which can be used instead of the diameter b to determine the scour for a group of two piles or more (Equation 7.28 of HEC-18):

$$a_{pg}^* = a_{proj} K_{sp} K_m \qquad (9.7)$$

where a_{proj} is the sum of the non-overlapping projected widths of piles, K_{sp} is a coefficient for pile spacing and K_m is a coefficient for the number of aligned rows, m (see Figure 9.20 for more clarification).

For a flow angle of 0°, the projected width in our case study is simply three times the diameter b. In case of skewed flow, this projected width can increase significantly. Including all piles from a pile group when determining the projected width results in an overestimation as the impact of the piles in the last row weighs as much as those in the front. Attempts by Smith (1999) to derive weighting factors did not prove worthwhile. A reasonable alternative is to only include the first column and the first two rows, as illustrated in Figure 9.20 (figure 7.10 from HEC-18), which in the case of the pile groups in the Scheldt means that the projected width can vary from three to

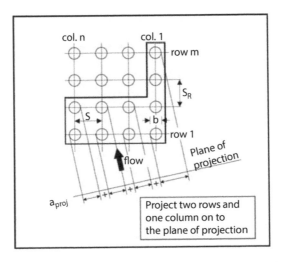

Figure 9.20 Projected width for skewed flow (figure 7.10 from HEC-18).

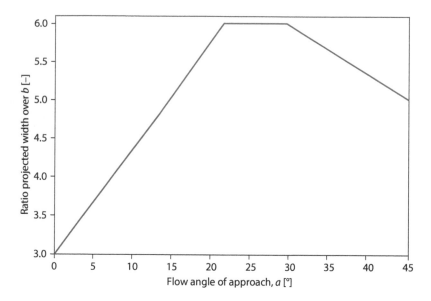

Figure 9.21 Dependence of the projected width on angle of approach flow.

six times the diameter b (denoted with 'a' in Figure 9.20), depending on the flow angle. This dependence is shown in Figure 9.21.

In the case of a group of non-cylindrical piles and a non-zero angle of attack, including both $K_\omega > 1$ for the angle of attack and a projected width as described above would be equivalent to accounting for flow skewedness twice. Therefore, it is advised to omit K_ω for pile group calculations.

As the pile groups in the Scheldt are aligned in the streamwise direction, a projected width of three times the diameter is sufficient.

The pile spacing factor K_{sp} and the factor for the number of aligned rows K_m both follow from equations of Sheppard and Glasser (2004). K_{sp} depends on the ratio between the single-pile diameter b and the pile spacing S, as well as on the ratio between the single-pile diameter b and the projected width a_{proj}:

$$K_{sp} = 1 - (4/3)\left(1 - b/a_{proj}\right)\left(1 - (S/b)^{-0.6}\right) \tag{9.8}$$

K_m is 1 if the skew angle is greater than 5 degrees. For more than five rows or piles aligned in the direction of the flow, K_m is 1.19. In this case, with parallel flow and three aligned rows, the following equation holds

$$K_m = 0.045m + 0.96,$$

with m the number of aligned rows

This leads to $K_{sp} = 0.62$ and $K_m = 1.095$ and thus to an equivalent width a^*_{pg} of 2.18 m to be used instead of diameter b in Equation (8.10) (with K_i still equal to 1.1).

Table 9.9 Results of scour hole calculations

Location_pile group	$y_{m,e}$ for single pile (m)	$y_{m,e}/b$ (–)	$y_{m,e}$ for group of piles (m)	$y_{m,e}/a^*_{pg}$ (–)
1_1/2	1.10	1.0	1.75	0.8
1_3/4	1.24	1.2	1.98	0.9
2_1/2	1.72	1.6	2.74	1.3
2_3/4	1.60	1.5	2.55	1.2

9.6.4 Results and discussion

The results are shown in Table 9.9. The equivalent width a^*_{pg} of 2.18 m leads to a constant factor of 1.59 between the scour depths for a group of piles and a single pile (compare columns 2 and 4 in Table 9.9).

The effective width of the pile group a^*_{pg} can also be used to determine the required dimensions of the scour protection in the horizontal plane. The distance between the four pile groups per pylon (S2) is 26.4 m, which is over 11 times a^*_{pg}. Therefore, no influence between the four pile groups is to be expected as the required dimensions for scour protection do not overlap.

However, there is a difference between the area for which the influence of a pile or pile group is felt and the area which is eventually affected by scour. For example: an analysis of the bed topography showed that the silty bottom lies stable under an angle of 1V:15H. This means that a scour hole with a depth of 2.5 m will eventually have horizontal dimensions of approximately 75 m, meaning that the scour holes of the different pile groups will overlap. Based on these findings a new bottom profile was generated for which the stability of the river dikes was evaluated (see Figure 9.22). This resulted in the necessity to design a bottom protection, as the river dikes were no longer stable in case of the eroded bottom profiles.

9.7 Scour development in front of culvert (van Oord)

9.7.1 Introduction

This case addresses the strengthening of the protection downstream of a culvert in a tidal area. It was observed that scouring downstream of the initial protection endangered the stability of the culvert. Therefore, an improved design was made and built. In this Section, the scope is described. Section 9.7.2 describes the initial bed protection and the monitored scour hole development. Section 9.7.3 presents the new design of the bed protection consisting of concrete blocks. The blocks were intended to decrease the flow velocities and to spread the flow in order to minimise the scour depth downstream. Section 9.7.4 presents the results of the redesign.

For a project in the province of Zeeland, The Netherlands, a culvert was installed through a dike enabling natural tidal action to occur within the polder hinterland. It had been determined that flow discharge rates of up to 78 m³/s could occur through the culvert during ebb. The culvert consisted of three rectangular sections, each 3.5 m wide and 2.1 m high, giving a total cross-sectional area of width × height = 10.5 m × 2.1 m.

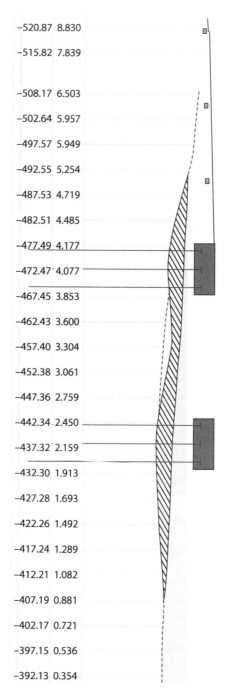

-520.87 8.830

-515.82 7.839

-508.17 6.503

-502.64 5.957

-497.57 5.949

-492.55 5.254

-487.53 4.719

-482.51 4.485

-477.49 4.177

-472.47 4.077

-467.45 3.853

-462.43 3.600

-457.40 3.304

-452.38 3.061

-447.36 2.759

-442.34 2.450

-437.32 2.159

-432.30 1.913

-427.28 1.693

-422.26 1.492

-417.24 1.289

-412.21 1.082

-407.19 0.881

-402.17 0.721

-397.15 0.536

-392.13 0.354

Figure 9.22 **New bottom profile.**

Figure 9.23 **Overview of the culvert: lower left is the seaward side, upper right is the polder hinterland.**

The bottom level of the culvert was situated at a level of −1.6 m +NAP. This resulted in maximum flow velocities of about 4 m/s at the exit of the culvert. Under these conditions, the flow is supercritical, and a highly turbulent jet occurs.

The area at the seaward side consisted of two 13 m long sheet-pile guide walls at an angle of 30° to the main flow direction. Between the guide walls the bottom protection consisted of 10–60 kg rock on a geotextile and fixed by colloidal concrete at −2.0 m +NAP. A short transition slope of 1:10 connected the bottom protection to the culvert. Due to requirements for water levels in the hinterland, a typical stilling basin could not be applied here. The bed material consisted of fine sand (d_{50} = 130 μm).

Shortly after commissioning of the facility a scour hole developed at the end of the bottom protection, more rapidly than envisaged. The causes were evaluated. A redesign of the bed protection indicated that a length of 200 m would be required to limit the equilibrium depth of the scour hole to 5 m. The present case study highlights the redesign process and the design concept finally chosen.

Figure 9.23 shows an aerial photograph of the project prior to opening, at a moment when the water levels on both sides of the culvert were about equal. The area to the lower left is the seaward side and the upper right is the nature area to be inundated later. The curved dam on the inner side was installed to maintain a minimum water level in the area.

9.7.2 Initial bottom protection and scouring

The initial bottom protection on the seaward side consisted of a 20 m long horizontal bed at −2 m +NAP followed by a 30 m section at a 1:10 slope with the seaward end at −5 m +NAP. The bed protection consisted of 10–60 kg rock. The first 20 m section was penetrated with colloidal concrete, and the outer 30 m section was loose rock on a thick granular filter. The design was made after the flow had been modelled numerically and by applying the scour formula of Dietz (1969). The flow jet was represented as a wall jet and assumed to have 6° lateral diffusion, starting at a distance of 21 m from the culvert opening. Applying a relative turbulence intensity $r_0 = 0.2$ led to a (maximum) turbulence coefficient of $\omega = 1+3r_0 = 1.6$ in the Dietz formula. For a water depth of 2.7 m, an average current speed of 1.5 m/s at the end of the scour protection, a critical flow velocity of 0.4 m/s, and the expected equilibrium scour depth was calculated to be 13.5 m.

The scour development was monitored after extending the bottom protection. In eight weeks, the scour hole had attained a depth of 11.5 m below the original bed level. An impression of the jet flow during ebb is shown in Figure 9.24. The piles are located 40 m from the culvert outlet.

The development of the scour hole is shown in Figure 9.25.

Figure 9.24 Jet during ebb flow into the tidal inlet. The poles are located 40 m from the culvert.

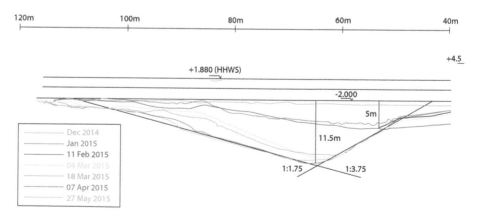

Figure 9.25 Cross-section of initial scour protection and the scour hole development.

The profile measurements were analysed and compared to the various prediction methods. This resulted in equilibrium scour depths from 11 m with Equation (4.14) in the previous scour manual (Hoffmans & Verheij, 1997) to more than 100 m with Equation (5.18). It was found that the combination of the Dietz formula Equation (3.16) with the relative turbulence intensity from Equation (5.17) compared well to the measurements, as the protection length was relatively long (50 m) compared to the water depth (varying between 1 and 5 m). Furthermore, the situation was one of clear-water scour because the scour protection on the inland side of the culvert prevented erosion of sediment. The measured scour development in time with Equation (6.13) was reproduced well when applying the 2D turbulence coefficient from De Graauw and Pilarczyk for a value of $\alpha = 2.6$. This led to an equilibrium depth of about 16 m, when taking the effect of tidal action into account through a coefficient $\eta = 0.8$ to compute the characteristic mean velocity with Equation (6.15).

9.7.3 New design bottom protection

The aforementioned methods for computing the length of protection required to limit the scour depth to 5 m, led to a length of more than 200 m. This was not practical for several reasons and a different concept was thus required to reduce the flow velocities and turbulence. After evaluating different concepts, it was decided to not remove or modify the structure close to the opening of the culvert to reduce the risk of damaging the culvert or the primary sea dike. A pile screen concept was therefore eliminated since that would require modifications to the bottom protection and fixations to the guide walls. The chosen concept was then a "block screen" barrier, consisting of a few concrete blocks placed on top of the existing bed some distance before the opening. Simplified flume testing showed that a configuration of six blocks placed in a shallow arc were effective in dissipating and spreading the flow jet. The total blocking coefficient of this configuration was about 50%, which leads to flow velocities between the blocks of about twice those at the outlet. The bed protection was penetrated with colloidal concrete at that location to ensure adequate stability. The blocks needed to have a mass of about 45 tonnes to remain stable during maximum flow conditions.

Computations for the flow dispersion were then made along with sensitivity analyses for the degree of spreading of the flow, the loss coefficient across the block screen, the turbulence intensity and the equilibrium scour depth. These computations were also compared to model tests performed in an earlier investigation with a pile screen (Delft Hydraulics, 1982). From this investigation we obtained valuable information about the influence of blocking percentage and location on the barrier's hydraulic resistance and the distribution of flow behind the barrier. The energy loss over the barrier is described by $\Delta H = \xi \cdot u^2/2g$ in which the loss coefficient ξ was approximated to be between 0.6 and 1.5 at ebb flow. The average current speed was then computed at various distances from the block screen for various values of lateral jet spreading. A final length of 100 m from the culvert was chosen for the bed protection. At that point, the current speed was estimated to be less than 0.5 m/s and the resulting equilibrium scour depth less than 5 m.

In total, six concrete blocks were installed at a distance of 15 m in front of the outlet. In addition, the guide walls were 'extended' by placing two more concrete blocks at the ends of the walls. A plan view of the resulting block screen barrier is shown in Figure 9.26. The dominant loading condition occurs during ebb flow, when the current exits the culvert (lower part of the figure) with a velocity of about 4 m/s (flowing toward the top of Figure 9.26).

Once the flow behind the block screen was determined, the extent of the bottom protection needed to be determined. The resulting scour hole was required not to exceed 5 m depth. As in the previous design, the formula from Dietz (1969) was applied to calculate the equilibrium scour depth. Slightly conservative values of the various coefficients were chosen, and a sensitivity analysis was performed later. A turbulence coefficient of $\omega = 1.75$ was applied, resulting from a relative turbulence intensity factor of $r_0 = 0.25$ that was determined for a 2D culvert.

The bottom protection followed a 1:10 slope, and the flow was assumed to be completely distributed over the depth at the end of the bottom protection. The angle of lateral spreading of the flow was assumed to vary from 1:8 (optimistic) to 1:30 (conservative). The sensitivity to various input parameters was analysed, applying 'optimistic' and 'pessimistic' assumptions. This yielded a trade-off matrix and resulted in bottom protection lengths ranging from 80 to 115 m. Eventually a length of 100 m was chosen. The flow velocity at that distance was estimated to be below 0.5 m/s. The lateral extent of the bottom protection was chosen to follow a 1:6 lateral angle. The resulting design is shown in Figures 9.27 and 9.28.

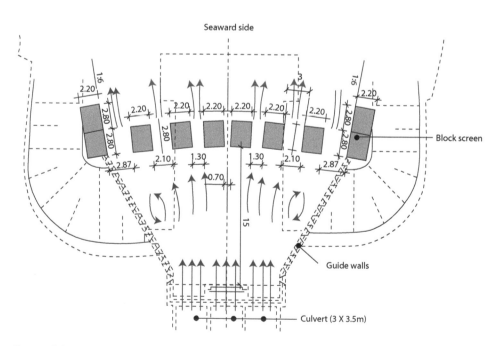

Figure 9.26 **Plan view of the 'block screen' installed at the seaward outlet of the culvert.**

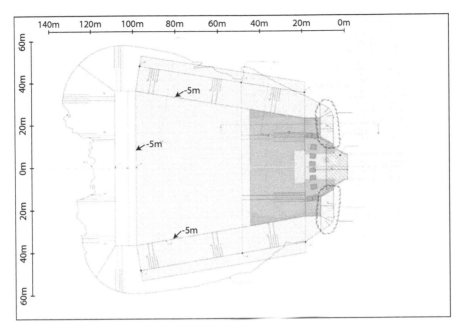

Figure 9.27 **Plan view of the final design, showing the block screen at the culvert outlet and the bottom protection in dark grey (penetrated) and light grey (loose rock).**

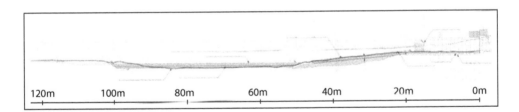

Figure 9.28 **Cross-section of the final design.**

9.7.4 Result redesign

One year after re-commissioning, the scour development was monitored and shown to be less than 1m deep, in line with the expectations. As of October 2018, no further issues have been observed.

9.8 Bed protection at railway bridge in a bypass of the river Waal (Deltares)

9.8.1 Introduction

This case addresses the bed protection at a railway bridge over the Spiegelwaal river bypass at Nijmegen. The bypass conveys flows during floods and has been designed for floods with a probability of exceedance of 1 in 1250 years. The same design flood was

Figure 9.29 **Railway bridge over the bypass in Nijmegen.**

used for the bed protection. This section introduces the situation. The flow conditions are described in Section 9.8.2. The scour depth at the bridge piers if no bed protection would be present is computed in Section 9.8.3. Section 9.8.4 addresses the design of the bed protection with special attention to the required length for preventing shear failure. Section 9.8.5 gives some final remarks that include safety factors and failure probability.

The Spiegelwaal is a bypass channel of about 4 km length in the river Waal at Nijmegen. This channel was constructed between 2013 and 2015 under the Room-for-the-River programme. The railway bridge spans the main river and the bypass of the river Waal, connecting the city of Nijmegen and the Lent district north of Nijmegen.

Figure 9.29 shows the railway bridge with its three streamlined piers. Usually the water in the Spiegelwaal is in rest. However, during floods this bypass conveys part of the discharge to the sea, reducing water levels and, hence, the risk of overflowing or failing of the main river dikes. The piers consist of a slender part on a footing. The width and the length of the footing are 16.5 m and 40.0 m, respectively, and the width and length of the piers above are 5 m and 15 m, respectively. The width of the bypass is 170 m.

9.8.2 Flow condition

According to the flow model WAQUA, the depth-averaged flow velocities are for design conditions about 2.0 m/s (flow depth is 13.5 m). Between the footings of the piers, the mean velocities were determined analytically to be about 2.5 m/s (Straathof, 2015). The total width of the three footings (about 50 m) influences the velocity in the constriction enormously.

9.8.3 Scouring

The equilibrium scour depth at the footing of the pier ($b = 16.5\,\text{m}$) can be estimated with Equation (7.12) (see also Section 8.6.3 and Figure 8.14):

$$y_{m,e} = h_0\left[(1-b/B)^{-\frac{2}{3}} - 1\right] + K_B b\tanh\left(h_0/b\right) \tag{7.12}$$

$$y_{m,e} = 13.5\left[(1 - 50/170)^{-\frac{2}{3}} - 1\right] + 1.5 \cdot 16.5\tanh(13.5/16.5) = 3.5 + 16.5 = 20 \text{ m}$$

Hence, very large scour depths can be expected during floods; most likely even larger than 20 m since the alignment of the pier to the flow has been neglected here. For instance, the scour depth increases by a factor of about 1.3 if the angle between the main axis of the horizontal cross-section of the pier and the direction of the approach flow equals 15°. "Thus the scour depth is $1.3 \times 20 \text{ m} = 26 \text{ m}$".

9.8.4 Designed bed protection

According to Bonasoundas (1973; see also Figure 8.14), the width and length to be protected at a circular pier are six times and seven times the footing width, respectively, and thus the area to be protected is $100 \text{ m} \times 115 \text{ m} = 115{,}000 \text{ m}^2$. Since there is an overlap in the width direction the complete river bed should be protected (Figure 9.30). Upstream of the bridge pier the length of the bed protection is three times the footing width ($= 3 \times 16.5 = 50\,\text{m}$) and downstream four times ($= 4 \times 16.5 = 66\,\text{m}$).

Since the bed is protected, the bed turbulence is larger than for uniform flow conditions and thus scour can be expected at the transition of the bed protection (stone weight 5–40 kg). By using $r_0 = 0.2$, the maximum scour depth equals the flow depth of 13.5 m after 9–10 days ($t_1 = 220\,\text{h}$), which is about the duration of a flood. Hence, the ratio between the bed protection and the scour depth is $L/y_{m,e} = 65/13.5 = 4.8$, which is satisfactory for preventing a shear failure since a minimum of 3 is required (see also Table 3.4). However, no safety factor has been applied and the computed scour depth is the expected average scour depth. The criteria in Table 4.4 are valid for average scour depths. Moreover, Figure 6.11 shows that the upstream scour slope will be unstable if $U_0 = 2\,\text{m/s}$ and $r_0 = 0.2$. Thus, even for the average scour depth of 13.5 m the situation is unstable.

The bypass was designed for high discharges and then an extreme scour depth might occur of two times 13.5 m. This results without safety factor in a ratio $L/y_{m,e} = 65/(2 \times 13.5) = 2.4$, which means that a shear failure cannot be excluded.

9.8.5 Final remarks

Summarising, the formulas in this manual result in a design that is critical for the stability of the upstream scour slope. Therefore, the dimensions of the scour hole need to be monitored after a flood. During the design, the 1997 version of this scour manual was used, which did not contain a safety approach based on statistics with an average value and a standard deviation. The present scour manual does address a safety approach. Surely, using an extreme scour depth results in a conservative approach, but

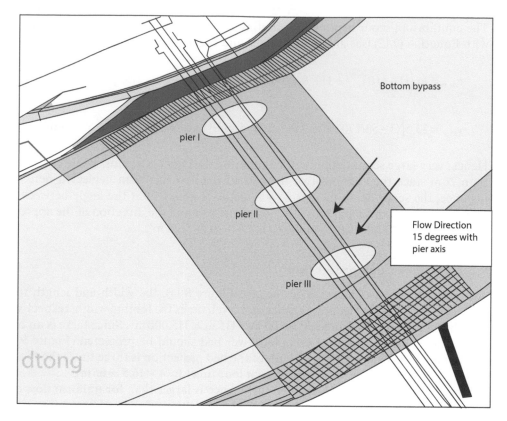

Figure 9.30 **Overview of the bed protection near the railway bridge at Nijmegen (river flows from right to left).**

as yet no safety factors are available as a function of shear failure probability. Based on the ratio of $L/y_{m,e} = 3.0$, a safety factor can be derived from $65/(\gamma_s \times 13.5) = 3.0$. This yields $\gamma_s = 1.6$. According to Johnson (1992), who derived failure probabilities for bridge piers, this means an annual probability of failure of 0.001 (see Table 2.2), which is close to the failure probability of 1/1250 per year for the occurrence of the 16,000 m³/s design flood.

9.9 Pressure scour around bridge piers (Arcadis)

9.9.1 Introduction

Scour in this case is related to water flowing against or over a bridge deck. The case presents how to compute pressure scour. Basically, the bridge deck obstructs the flow of water. This section introduces the problem, whereas Section 9.9.2 presents the flow conditions. In Section 9.9.3, the pressure scour formulas are presented. In Section 9.9.4, the computed scour results are presented and discussed.

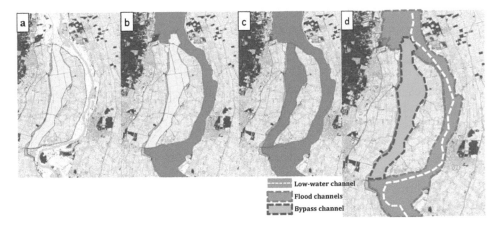

Low-water channel
Flood channels
Bypass channel

Figure 9.31 Inundated areas during low discharge (a), high discharge (b), an extreme flood (c) and main river and flood bypass (d).

The flood bypass between Zwolle and Deventer (the Netherlands) alongside the river IJssel was realised in 2017. It conveys water at extreme floods with an estimated frequency of once every 100 years (see Figure 9.31). This reduces upstream flood levels up to 0.7 m. The bypass is part of the Room-for-the-River programme that increases flood protection and environmental conditions along the Rhine branches in the Netherlands. Arcadis acted as a designer in many of the Room-for-the-River projects.

Several small bridges have been built over the bypass. These bridges assure the accessibility and preserve the agricultural function of the area. When opening the flood gates during a flood event, these bridges will gradually submerge. The bridges will restrict the flow. This funnelling of water leads to pressure scour underneath and downstream of the bridge. This is considered in this case.

9.9.2 Flow conditions

We consider the bridge 'K07 Breeweg' (see Figure 9.32) as an example. The following conditions apply to this bridge:

- The construction has a total span of 20 m, which is supported in the middle by a single bridge pier. The bridge pier consists of three foundation piles ($450 \times 450\,mm^2$) and a girder, parallel to the direction of the channel. The deck height is 0.7 m.
- The distance between the bridge deck and the bed of the channel is 2.3 m.

The main factors for scour at and around bridges are the water depth h_u and the flow velocity U. In total, five situations were examined, disregarding intermediate situations.

- Case 0a: Low discharge, no scour
  The water depth is regulated at 1.3 m during the winter period. The maximum velocity is determined to be 0.3 m/s, for which no scour is expected to occur.

- Case 0b: Flood, no *pressure* scour

   The channel gates have just been opened at the start of an extreme flood. The water level has almost reached the bottom of the bridge deck, which is 2.3 m above the upstream bed. The corresponding maximum velocity U of 0.5 m/s has been calculated using WAQUA.

In the two cases, above the presence of the bridge deck does not influence the flow of water. When the water rises even more and starts to interact with the bridge deck, the following cases can be distinguished (Figure 9.33):

- Case 1: Flood, bridge partially submerged, no eddies

   The water level in the channel is rising and reaches the top of the bridge deck at 2.9 m above the upstream bed. Flow velocity is 0.5 m/s.
- Case 2: Flood, bridge partially submerged, with eddies

   Equal to Case 1, but with turbulence around the bridge pier, as shown in Figure 9.32 in 'Case 2'.
- Case 3: Flood, bridge fully submerged

   The water has just submerged the bridge deck ($h_u = 3.1$ m). Flow velocity is 0.5 m/s.

Figure 9.33 (taken from Section 8.2.2) represents these three cases schematically. The rising water level on both sides of the bridge determines which case applies.

   The flow condition of Case 1 represents only a short transition to Case 2. As scour below the bridge develops, the eroded materials will be deposited downstream of the bridge. This amount of sediment might raise the tailwater and cause the downstream deck to become submerged and tail controlled. As the duration of Case 1 is short, and not of interest for the phenomenon of pressure scour, we do not consider this case any further.

Figure 9.32 **Bridge 'K07 Breeweg'.**

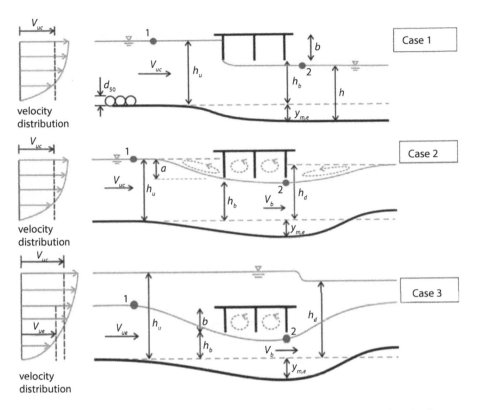

Figure 9.33 Pressure scour cases occurring underneath a flooded bridge deck.

9.9.3 Scour computation

We assumed a median sand diameter d_{50} of 250 μm. Scour starts when the bottom material becomes mobile (clear-water scour). This initiation of motion condition is determined by calculating a critical velocity V_{uc}. We applied the formula below to determine the critical velocity:

$$V_{uc} = 7.69 \cdot \sqrt{\psi_c (s-1) g d_{50}} \cdot \left(\frac{h_u}{d_{50}} \right)^{1/6}$$
(8.19)

where
V_{uc} = Critical velocity (m/s),
Ψ_c = Critical Shields parameter (–),
Ψ_c = 0.043 (–) (see Table 4.4),
s = specific density of bottom material (-),
d_{50} = median diameter (m),
h_u = upstream water depth (m).

In case a formula with the Manning coefficient n is preferred, the following formula can be used:

$$V_{uc} = \frac{\sqrt{\psi_c (s-1) d_{50}}}{n} \cdot h_u^{1/6} \tag{9.8}$$

with $n = 0.04 \left(1.25 \cdot d_{50}\right)^{1/6}$ $(s/m^{1/3})$

In Case 3, an effective velocity V_{ue} is applied because the water flows over the top of the bridge deck. The formula for V_{ue} reads:

$$V_{ue} = V_{uc} \left(\frac{h_b + a}{h_u}\right)^{0.85} \tag{8.21}$$

where
h_b = Water depth under bridge (m)
a = Deck block height (m)

Section 8.4.2 presents formulae to calculate the equilibrium depth of the pressure scour hole:

$$y_{m,e} \left(Fr_s, h_b, a\right) = \left(h_b + a\right) \sqrt{\frac{1 + 1.3680 / Fr_s^{2.409}}{1 + 1.8652 / Fr_s^2}} - h_b \tag{8.17}$$

with

$$Fr_s \left(V_{uc}, h_u, h_b\right) = \frac{V_{uc}}{\sqrt{g \left(h_u - h_b\right)}} \tag{8.18}$$

Figure 8.9 shows Equation (8.17) in another form:

$$Z = \frac{h_b + y_m}{h_b + a} = \sqrt{\frac{1 + 1.3680/Fr_s^{2.409}}{1 + 1.8652/Fr_s^2}} \tag{8.20}$$

where Z = scour number (–).

The scour number Z is a relation between water depth below an obstacle, deck block height and scour depth which was derived from experimental tests. The Froude number relates critical flow velocity (V_{uc} or V_{ue}), upstream water depth h_u and water height under the bridge h_b.

9.9.4 Results

The values for bridge 'K07 Breeweg' in Table 9.10 were derived for Cases 2 and 3. Case 1 is only temporary and not considered any further.

The computational results are given in Table 9.11. The pressure scour depth Ys is given in the 4th column. The 'normal' scour depth around a bridge pier (Equation 8.4) is given in the 5th column. The difference in scour depth is small due to the used depths of 2.9 m for Case 2 and 3.1 m for Case 3.

Table 9.10 **Flow conditions**

Situation	Water depth under bridge h_b (m)	Water depth h_u (m)	Deck block height a (m)	Max. depth averaged flow velocity (upstream) V_{up} (m/s)	Critical velocity (calculated) V_{uc} (m/s)	Critical effective velocity (calculated) V_{ue} (m/s)
Case 2	2.3	2.9	0.6	0.35	0.48	n/a
Case 3	2.3	3.1	0.8	0.35	0.49	0.48

Table 9.11 **Scour depths**

Situation	Water level (m)	Critical velocity (m/s)	Froude number Fr (–)	Scour depth Ys (m)	Scour depth[a] $Y_{m,e}$ (m)
Case 2	2.9	0.48	0.20	1.15	1.45
Case 3	3.1	0.49	0.17	1.51	1.48

[a] Scour depth according to Equation (8.4), circular pier, width pier = height bridge deck (b)

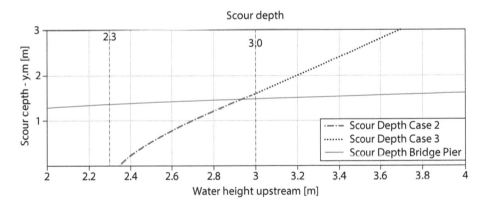

Figure 9.34 **Water level versus scour depth.**

Figure 9.34 presents the scour depth for Case 2, Case 3 and around bridge piers for a range of upstream water levels. The bottom of the bridge deck is located at 2.3 m above the bed and the top of the bridge deck is located at 3.0 m above the bed. These values are indicated by red vertical lines. The flow condition corresponding to Case 2 is valid up to an upstream water depth of 3.0 m. Case 3 applies when the water depth is higher dan 3.0 m. The scour depth increases and is significantly larger than the expected scour around bridge piers for an increasing water level. In this example, the scour depth is larger than the scour around bridge piers only when Case 3 applies.

9.10 Bed protection at the weir at Grave in the river Meuse (Rijkswaterstaat)

9.10.1 Introduction

The case addresses the scour due to a damaged weir as well as the design of the new bed protection. The primary attention of the case focusses on measures to control further damage. Section 9.10.2 describes the situation and the scouring directly after the accident due to a drifted ship. Section 9.10.3 presents the flow condition that may occur in winter during high discharges which were computed with a CFD model. In Section 9.10.4, the scour and a temporary bed protection were designed for an expected high discharge with particular attention to the extent of the protection regarding shear failures. Section 9.10.5 describes the situation after a flood and the effect of the temporary protection. Finally, Section 9.10.6 describes a hindcast which was carried out after the flood using a CFD model and an analytical formula.

9.10.2 Scope

The river Maas, situated in the southeast of the Netherlands, has regulated water levels by seven weirs, combined with ship locks to make the rain-fed river navigable. One of these is at Grave, a weir combined with a bridge and built in 1929. The neighbouring weirs are at Sambeek, upstream, and at Lith, downstream. Figure 9.35 shows a cross-section of the weir at Grave. Downstream of the weir panels, a stilling basin reduces the energy in the flow and concrete blocks protect the bed between the bridge piers. Further downstream the bed consisted of 23 m grouted rock followed by the unprotected sandy bed.

In the evening of 29 December 2016, in thick fog and darkness, the Maria Valentine, a 2000 ton benzene inland tanker, missed the ship lock, collided with the eastern opening of the weir, went through it and damaged five support beams (see Figure 9.36).

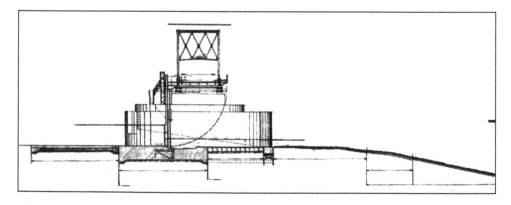

Figure 9.35 Cross-section of weir at Grave.

Figure 9.36 **Damaged weir at Grave.**

As a result, the upstream impounded reservoir was emptied into the downstream reservoir, rendering navigation and other river-bound activities (such as ferries, living boats and quay use by local business) impossible. Luckily the river discharge was low (approximately $100\,m^3/s$) and therefore the concentrated flow through the gap caused only minor erosion downstream (see Figures 9.37 and 9.38).

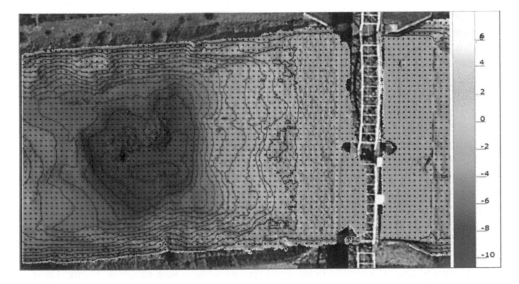

Figure 9.37 **Survey 4-01-2017 after the calamity (coloured depth scale in metres NAP).**

Figure 9.38 Erosion (dark and light blue) and sedimentation (red and yellow) caused by the concentrated flow through the gap (arrow indicates the location of the gap).

9.10.3 Flow condition

The primary task was to restore the water level upstream. This was accomplished by the construction of a 5.5 m high rock dam with impermeable core downstream of the eastern opening. Initially, the entire bed protection downstream of the weir was extended to 50 m to protect against the higher flow velocities in the closure gap of the dam but also against the increased velocities in the western opening when the dam was completed (the discharge opening of the weir was reduced by more than 50%). Later, a more detailed analysis (de Loor & Weiler, 2016) showed that in the range from 400 till 1000 m³/s the flow through the reduced opening would be supercritical and that the stilling basin was too shallow and too short to contain the hydraulic jump. The highly erosive hydraulic jump would be situated above the grouted bed protection and the sandy bed. For example, at a discharge of 840 m³/s the head difference over the weir is 1.7 m and the velocity at the weir is 6.9 m/s, with a Froude number equal to 1.27. With this Froude number, the jump will not be compact. The velocities are illustrated by the result of a CFD computation in Figure 9.39.

In the winter season, a discharge of 1000 m³/s is likely to occur and therefore the length and strength of the bed protection had to be designed for this discharge. Without additional measures, the turbulent jump would occur on top of the grouted bed protection near the bridge piers. The turbulent pressure fluctuation is defined in N/m² and related to the differential head, thus recomputed into a range of 1–2 m. The present grouted bed protection is only 0.7 m thick and possibly insufficient to withstand these pressure fluctuations. Unfortunately, at the beginning of March already a flood of 900 m³/s was expected, and a robust measure had to be taken quickly.

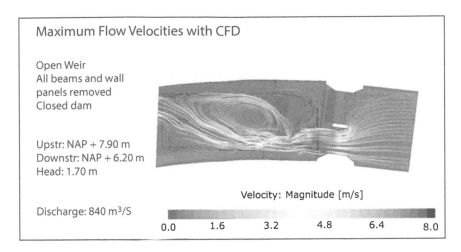

Figure 9.39 **Velocities around weir from CFD computation.**

9.10.4 Scour and bed protection

As a countermeasure, the bed between the bridge piers was ballasted with the heaviest rock stones of a 3–6 tonnes grading, placed in a single layer and in a formation to create extra stability against the 7 m/s flow. The stability of the rock was validated with Pilarczyk's stability formula. Downstream of this protection 4-tonne-rock nets were placed on top of the grouted bed protection. The protecting layer had to be thin since it would increase the velocities above and behind the layer.

Turbulence levels (r_0) behind the new sill would be higher than normal and were conservatively assumed 0.25. The equilibrium scour depth can then be estimated with Equation (5.6) for live-bed scour using the water depth h_0 for the length scale:

$$y_{m,e} = h_0 \cdot (1.0\,\chi - 1), \quad \text{with} \quad \chi = \frac{1 + 6.3 r_0^2}{1 - 6.3 r_{0,m}^2} \qquad \text{(6.4 and 6.5)}$$

The equilibrium scour depth is reached very quickly with these high flow velocity and turbulence levels.

With a downstream water depth $h_0 = 5.5$ m and assuming $r_{0,m} = 0.35$ as a safe estimate (underneath the hydraulic jump), the equilibrium scour depth is 28 m. Given the equilibrium scour depth a conservative estimate for the length of the bed protection can be made with Equation (3.19):

$$L = \frac{1}{2} y_{m,e} \left(\cot \gamma_2 - \cot \beta_a \right), \qquad \text{(3.19)}$$

with β_a and γ_2 the upstream slope angles before and after instability, respectively. With $\cot \beta_a = 2$ as a conservative estimate and $\cot \gamma_2 = 8$ (shear failure for coarse sand), the required length of the bed protection is 84 m, which is more than the present bed protection of 50 m long. The bed protection was extended by 40 m with 40–200 kg

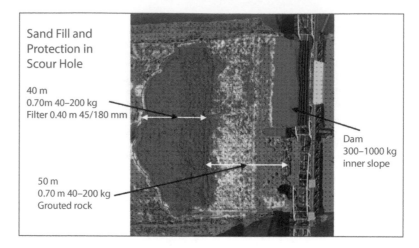

Figure 9.40 **Sand fill and protection before the flood.**

rock on a filter of 45/180 mm gravel. Where necessary the sandy slope was flattened with sand to 1:3, before adding the protection. The extended bed protection is shown in Figure 9.40.

9.10.5 Condition after the flood

The flood lasted about 10 days and reached a maximum discharge of 840 m³/s. During the flood a bank slide occurred on the west bank at the end of the bed protection (see Figure 9.41), probably caused by the deep scour hole. After the passage of the flood the

Figure 9.41 **Bank slide at the end of the bed protection.**

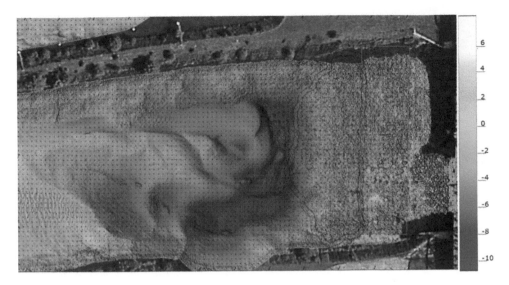

Figure 9.42 Scour depth after passage of the flood; survey 15-03-2017 (coloured depth scale in metres NAP).

bed protection was inspected and found to show only minor damage in the grouted section. This was repaired quickly to prevent further damage for the rest of the flood season. The west bank was also stabilised with a rock protection. Figure 9.42 shows the scour caused by the flood. The maximum scour depth is approximately 9 m and less than the predicted range. Due to the bank slide into the scour hole and the strong erosion in the eddy behind the dam, there was still a large quantity of sand in the system. Without this, the scour hole would have been deeper and larger. Inspection also showed that the heavy rock protection between the bridge piers had remained stable during the flood and that the rock-filled nets almost all had disappeared. The remainder of the flood season had no more floods and the repair works of the weir were completed before the next flood season.

9.10.6 Hindcast

CFD computations were made for this flood situation to determine the existing loads on the bed protection (O'Mahoney, 2018). This revealed turbulence intensities $r_0 = 0.15$ and $r_{0,m} = 0.3$, where r_0 was taken at the section where the bottom starts to deepen and the flow is still rather uniform (section D in Figure 9.43). Furthermore, $r_{0,m}$ was taken at the deepest point of the scour hole (section F in Figure 9.43). With Equation (6.5) this results in a scour depth of 9 m, which is the same as observed.

In another simplified approach the clear-water equilibrium scour depth was used and corrected for upstream sediment supply to get an estimate for the reduced scour depth (see 5.3.7). Since the Hoffmans formula (Equation 6.4) resulted in unrealistic scour depths, what could be expected because the formula is valid for clear water scour, the Dietz formula (Equation 3.16) was used to calculate the equilibrium scour depth:

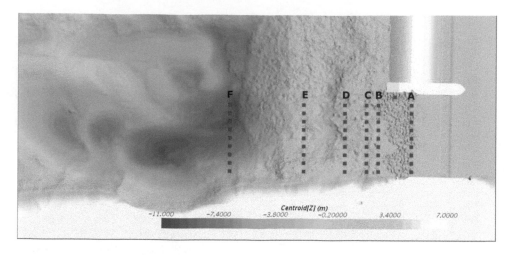

Figure 9.43 Output sections of CFD computation (O'Mahoney, 2018).

$$\frac{y_{m,e}}{h_0} = \frac{\omega U_0 - U_c}{U_c},\qquad(3.16)$$

with U_c = critical mean velocity (m/s), ω = turbulence coefficient and U_0 = mean velocity (m/s). According to Dietz, the average value for the turbulence coefficient is $\omega = 2/3 + 2r_0$, here resulting in $\omega = 0.97$. For the coarse sand of the river Maas ($d_{50} = 0.4$–2 mm), U_c ranges from 0.4 to 0.6 m/s (see Table 5.19 in CIRIA/CUR/CET-MEF, 2007). In the CFD computation, the average flow velocity U_0 is approximately 6 m/s at section D (O'Mahoney, 2018). Then, Equation (3.16) for the clear-water equilibrium scour depth ($y_{m,e,CL}$) results in a range of 48–74 m.

An alternative is to compute the live-bed equilibrium scour depth ($y_{m,e,LB}$) using Equation (6.17). First, we compute the volume of the scour hole (V_{CL}) with $c_a = 5$ for clear-water scour. Using this volume in the same equation with $c_a = 22$ for live-bed scour, the live-bed equilibrium scour depth ($y_{m,e,LB}$) can be calculated as

$$y_{m,e,LB} = \sqrt{\frac{V_{CL}}{22}} = \sqrt{\frac{5y^2_{m,e,CL}}{22}}$$

This results in a range of 23–35 m, which is far too high compared to observed.

Using data from CFD computations, the hindcast with the live-bed scour formula of Hoffmans gives a good prediction despite the amount of recirculating sand in the system. Clear-water scour computations largely overestimate the equilibrium scour depth. Also, the alternative method based on the volume of the scour hole overestimates the maximum scour depth.

References

Adnan, Z., & Wijeyesekera, D.C., 2007, Geotechnical challenges with Malaysian peat, *Proceedings of the Advances Computing and Technology 2007. Conference*, 23 January 2007, ed. Stephen et al., 252–261. London.W

Adnan, Z., Wijeyesekera, D.C., & Masirin, M.I.M., 2007, Comparative study of Malaysian Peat and British Peat pertaining to geotechnical properties, *Proceedings of Sri Lankan Geotechnical Society's First International Conference on Soil and Rock Engineering*, 5–11 August 2007, ed. H.S.W. Pinnaduwa, Colombo.

Ahmad, M., 1953, Experiments on design and behaviour of spur dikes, *Proceedings of the IAHR Conference*, Minnesota, 145–159.

Akkerman, G.J., 1976, Stormvloedkering Oosterschelde, Werkgroep 8, Vormgeving en constructie van sluitgaten, WL8–59, Ontgrondingen rond putten bij gedeeltelijk verdedigde bodem, Report M1402, Delft Hydraulics, Delft.

Altinbilek, H.D., & Basmaci, Y., 1973, Localized scour at the downstream of outlet structures, *Proceedings of the 11th Congress on Large Dams*, Madrid, 105–121.

Annandale, G.W., 1995, Erodibility, *Journal of Hydraulic Research*, 33(4): 471–494.

Annandale, G.W., 2006, *Scour Technology, Mechanics and Engineering Practice*, McGraw-Hill, New York.

Ariëns, E.E., 1993, Relatie tussen ontgrondingen en steenstabiliteit van de toplaag, Technical University of Delft, Delft.

Ashmore, P., & Parker, G., 1983, Confluence scour in coarse braided streams, *Water Resources Research*, 19(2): 392–402.

ASTM, 1992, *Soil and Rock; Dimension Stone, Geosynthetics*, Volume 04.08, Philadelphia, PA.

ASTM, 2017, Standard test method for consolidated undrained direct simple shear testing of fine grain soils. Active Standard ASTM D6528-17, ASTM International, USA.

Baldi, G. et al., 1982, Design parameters for sands from CPT, *Proceedings 2nd European Symposium on Penetration Testing*, Amsterdam.

Berg, M.B., 2001, Ontgrondingskuil van krib bij km 900.330. Universiteit Utrecht and RWS, Report DWW-2001-102.

Bishop, A.W., 1954, The use of slip circle in the stability analysis of slopes, *Proceedings of the European Conference on Stability of Earth Slopes*, I S.1–12, Stockholm.

Blaisdell, F.W., & Anderson, C.L., 1989, Scour at cantilevered outlets: Plunge pool energy dissipator design criteria, ARS-76, US Department of Agriculture, Agricultural Research Service, Washington, D.C.

Blazejewski, R., 1991, Prediction of local scour in cohesionless soil downstream of out-let structures, Agricultural University of Poznan, Faculty of Reclamation and Environmental Engineering, Poznan, Poland.

Blokland, T., 2013, Field test Parkkade. Scour by propeller jets of inland vessels against quay (In Dutch), Municipality of Rotterdam, Engineering Department, project HH2841, Rotterdam, the Netherlands.

Bogardi, J.L., 1974, *Sediment Transport in Alluvial Streams*, Akademiai Kiado, Budapest.

Bollaert, E., & Schleiss, A., 2003, Scour of rock due to the impact of plunging high velocity jets 1: A state of the art review. *Journal of Hydraulic Research*, 41(5): 451–464.

Bom, S., 2017, Scour holes in heterogeneous subsoil: A numerical study on hydrodynamical processes in the development of the scour holes, M.Sc. Thesis, Delft University of Technology, Delft.

Bonasoundas, M., 1973, Strömungsvorgang und Kolkproblem am runden Brückenpfeiler, Versuchsanstalt für Wasserbau, Bericht 28, Technischen Universität München, Oskar v. Miller Institut, Germany.

Bormann, N.E., & Julien, P.Y., 1991, Scour downstream of grade-control structures, *Journal of Hydraulic Engineering, ASCE*, 117(5): 579–594.

Brahms, A., 1767, *Anfangsgründe der Deich und Wasserbaukunst*, Aurich.

Breusers, H.N.C., 1966, Conformity and time scale in two-dimensional local scour, Proceedings of the Symposium on Model and Prototype Conformity, 1–8, Hydraulic Research Laboratory, Poona (also Delft Hydraulics, Delft, Publication 40).

Breusers, H.N.C., 1967, Time scale of two-dimensional local scour, *Proceedings of the 12th IAHR-Congress*, Fort Collins, Colorado, 275–282.

Breusers, H.N.C., 1971, Local scour near offshore structures, Symposium on *Offshore Hydrodynamics*, Wageningen, Netherlands (also Delft Hydraulics, Delft, Publication 105).

Breusers, H.N.C., Nicolet, G., & Shen, H.W., 1977, Local scour around cylindrical piers, *Journal of Hydraulic Research, IAHR*, 15(3): 211–252.

Breusers, H.N.C., & Raudkivi, A.J., 1991, *Scouring*, Balkema, Rotterdam.

Briaud, J.-L., 2008, Case histories in soil and rock erosion, *Fourth International Conference on Scour and Erosion*, Tokyo.

Briaud, J.-L., Chen, H.-C., Edge, B., Park, S., & Shah, A., 2001, *Guidelines for Bridges over Degrading and Migrating Streams*. Part 1: Synthesis of Existing (TX-01/2105-2), Transportation Research Board, Washington, DC.

Briaud, J.-L., 2019, personal communication.

Buchko, M.F., 1986, Investigation of local scour in cohesionless sediments by using a tunnel model, Report Q239, Delft Hydraulics, Delft.

Buchko, M.F., Kolkman, P.A., & Pilarczyk, K.W., 1987, Investigation of local scour in cohesionless sediments using a tunnel-model, IAHR-Congress, *Lausanne; Topics in Hydraulic Modelling*, 233–239.

Carstens, T., 1976, Seabed scour by currents near platforms, *3rd Conference on Port and Ocean Engineering under Arctic Conditions*, University of Alaska, 991–1006.

Chabert, J., & Engeldinger, P., 1956, *Étude des affouillements autour des piles de ponts*, Laboratoire National d'Hydraulique, Chatou.

CIRIA, 2002, *Manual on Scour at Bridges and other Hydraulic Structures* (1st edition). Report 551, London.

CIRIA, 2007, *Use of Vegetation in Civil Engineering*, Report C708, CIRIA, London.

CIRIA, 2015, *Manual on Scour at Bridges and Other Hydraulic Structures* (2nd edition). Report C742, London.

CIRIA/CUR/CETMEF, 2007, *The Rock Manual – The Use of Rock in Hydraulic Engineering* (2nd edition). Report CIRIA C683, London.

Clarke, F.R.W., 1962, The action of submerged jets on moveable material, *A Thesis Presented for the Diploma of Imperial College*, Department of Civil Engineering, Imperial College, London.

Compte-Bellot, G., 1963, Coefficients de dessymétrie et d'aplatissement, spectres et correlations en turbulence de conduite, *Journal de Mécanique*, 11(2): 105–128.

CUR 113, 2008, *Oeverstabiliteit bij zandwinputten*, CUR aanbeveling 113, Gouda, The Netherlands.

CUR 161, 1993, Filters in de waterbouw, SBRCURnet, Gouda, The Netherlands.

CUR 162, 1992, *Construeren met grond; Grondconstructies op en in weinig draagkrachtige en sterk samendrukbare ondergrond*. Ed. P. Lubking, Gouda, The Netherlands.

CUR, 1996, Erosie van onverdedigde oevers, CUR report 96–7, Gouda, The Netherlands.

CUR 233, 2010, Interface stability of granular filter structures, Theoretical design methods for currents. SBRCURnet, Gouda, The Netherlands.

CUR, 2015, Design formula for stable open granular filters loaded by currents, Kennispaper document K680.15, ISBN 978-90-5367-607-3, SBRCURnet, Delft, The Netherlands.

Dam, G., van der Wegen, M., Labeur, R.J., & Roelvink, D., 2016, Modeling centuries of estuarine morphodynamics in the Western Scheldt estuary, *Geophysical Research Letters*, 43, doi:10.1002/2015GL066725

Dargahi, B., 1987, Flow field and local scouring around a cylinder, Bulletin TITRAVBI-137, Department of Hydraulics Engineering, Royal Institute of Technology, Stockholm, Sweden.

de Graauw, A.F.F., 1981, Scour (in Dutch), Report M1001 Delft Hydraulics, Delft.

de Graauw, A.F.F., & Pilarczyk, K.W., 1981, Model-prototype conformity of local scour in non-cohesive sediments beneath overflow-dam, *19th IAHR-Congress*, New Delhi (also Delft Hydraulics, Delft, Publication 242).

de Groot, M.B. et al., 1988, The interaction between soil, water and bed or slope protection, *Proceedngs of the International Symposium on Modelling Soil-Water-Structure Interactions, SOWAS '88*, IAHR/ISSMFE/IUTAM, Delft (Ed. P. Kolkman et al).

de Groot, M., den Adel, H., Stoutjesdijk, T.P., & van Westenbrugge, K.J., 1992, Risk analysis of flow slides, *Proceedings 23rd International Conference on Coastal Engineering*, Venice.

de Groot, M.B., Van der Ruyt, M., Mastbergen, D., & Van den Ham, G.A., 2009, Bresvloeiing in zand, *Geotechniek*, 13(13): 34–39.

de Loor, A., Weiler, O., & Grave, W., 2016, *Temporary Situation after Collision (in Dutch)*, Report, Deltares.

de Vries, M., 1975, A morphological time scale for rivers, *Proceedings of the IAHR-Congress*, Sao Paulo, 17–23.

de Vries, M., 1993, River engineering, Lecture notes f10, Faculty of Civil Engineering, Hydraulic and Geotechnical Engineering Division, Delft University of Technology, Delft.

de Wit, L., 2006, Smoothed particle hydrodynamics, A study of the possibilities of SPH in hydraulic engineering, MSc thesis, Technical University Delft, http://resolver.tudelft.nl/uuid:eeb28db0-cc52-415c-932e-52ea4f9c9a86

de Wit, L., 2015, 3D CFD modelling of overflow dredging plumes. PhD thesis, Delft University of Technology, Delft.

Delft Hydraulics, 1972, *Systematic Investigation of Two and Three-Dimensional Local Scour (in Dutch)*, Investigation M648/M863/M847, Delft Hydraulics, Delft.

Delft Hydraulics, 1972a, Vergelijking van het stroombeeld en de uitschuring bij verschillende kribvormen, rapport M610, Delft.

Delft Hydraulics, 1972b, Vormgeving kribben voor oeverbescherming bij getijstromen, rapport M1032, Delft.

Delft Hydraulics, 1979, *Prototype scour hole 'Brouwersdam' (in Dutch)*, Investigation M1533, Parts I, II, III and IV, Delft Hydraulics, Delft.

Delft Hydraulics, 1986a, *Waterkrachtcentrale Linne*, Investigation Q243, Delft Hydraulics, Delft.

Delft Hydraulics, 1986b, *Waterkrachtcentrale Maurik*, Investigation M2110, Delft Hydraulics, Delft.

Delft Hydraulics, 1987, *Waterkrachtcentrale Alphen*, Investigation Q77, Delft Hydraulics, Delft.

Delft Hydraulics, 1988, *Ship-Induced Water Motions*, Rapport M1115, Delft.

Delft Hydraulics, 1989, *Stormvloedkering Nieuwe Waterweg*, Investigation Q965, Delft Hydraulics, Delft.

Delft Hydraulics, 1991, *Ontgronding bij een horizontale vernauwing*, Rapport Q935, Delft.

Delft Hydraulics, 2006, *Falling aprons*, Report Q4140, Delft

Deltares, 2015, *Opvullen en afdekken ontgrondingskuilen in de Oude Maas*, Rapport 1220038-015

den Haan, E.J., 1997, An overview of the mechanical behaviour of peats and organic soils and some appropriate construction techniques. In *Proceedings of Recent Advances in Soft Soil Engineering*, ed. Huat and Bahia, 85–108, Sarawak.

Diermanse, F.L.M., de Bruijn, K.M., & Beckers, J., 2015, Importance sampling for efficient modelling of hydraulic loads in the Rhine-Meuse delta. *Stochastic Environmental Research and Risk Assessment*, 29(3): 637–652.

Dietz, J.W., 1969, Kolkbildung in feinen oder leichten Sohlmaterialien bei strömendem Abfluß, Mitteilungen, Heft 155, Universität Fridericiana Karlsruhe.

Dietz, J.W., 1972, Ausbildung von langen Pfeilern bei Schräganströmung am Beispiel der BAB-Mainbrücke Eddersheim, *Mitteilungsblatt der BAW*, 31: 79–94.

Doehring, F.K., & Abt, S.R., 1994, Drop height influence on outlet scour, *Journal of Hydraulic Engineering, ASCE*, 120(12): 1470–1476.

Dorst, K., Meijs, D.J., Schroevers, M., & Verheij, H.J., 2016, Prototype measuring of erosion and currents under the keel of a sailing ship in a canal. *ICSE 2016, 8th Conference*, London.

Eggenberger, W., & Müller, R., 1944, Experimentelle und theoretische Untersuchungen über das Kolkproblem, Mitteilungen aus der Versuchsanstalt für Wasserbau 5, Zürich.

Einstein, H.A., 1950, The bed-load function for sediment transportation in open channel flows, Technical Bulletin 1026, US Department of Agriculture, Washington, D.C.

Ettema, R., 1980, *Scour at Bridge Piers*, Report 216, University of Auckland, Auckland.

Fahlbusch, F.E., 1994, Scour in rock riverbeds downstream of large dams, *The International Journal on Hydropower & Dams, IAHR*, 1(4): 30–32.

FAP21/22-RRI, 1993, Bank Protection and River Training (AFPM) Pilot Project, FAP21/22 (RRI) – Final report planning study (draft), Dhaka (also report Q1326 of Deltares, *Guidelines and design manual for standardized bank protection structures*)

Farhoudi, J., & Smith, K.V.H., 1982, Time scale for scour downstream of hydraulic jump, *Proceedings ASCE*, 108(HY10): 1147–1161.

Farhoudi, J., & Smith, K.V.H., 1985, Local scour profiles downstream of hydraulic jump, *Journal of Hydraulic Research, IAHR*, 23(4): 343–358.

Fellenius, W., 1947, *Erdstatische Berechnungen mit Reibung und Kohäsion und unter Annahme kreiszylindrischer Gleitflächen*, Ernst, Berlin.

FHWA, 1985, Design of spur type stream bank stabilization structures, Federal Highway Administration, Report FHWA-RD-84/101.

FHWA, 2009, Bridge pressure flow scour for clear water condition. Federal Highway Administration, Report FHWA-HRT-09-041.

FHWA, 2015, Scour in cohesive soils, Federal Highway Administration, Report 009463.

Froehlich, D.C., 1988, Analysis of onsite measurements of scour at piers, *Proceedings of the ASCE, National Hydraulic Engineering Conference*, Colorado Springs, Colorado.

Galay, V., 1987, Erosion and sedimentation in the Nepal Himalaya. Ministry of Water Resources of His Majesty's Government of Nepal.

Gingold, R.A., & Monaghan, J.J., 1977, Smoothed particle hydrodynamics: theory and application to non-spherical stars, *Monthly Notices of the Royal Astronomical Society*, 181(3): 375–389, https://doi.org/10.1093/mnras/181.3.375

Graf, W.H., 1971, *Hydraulics of Sediment Transport*, McGraw-Hill, New York.

Grotentraast, G.J., Bouman, W.J., Gelok, A.J., & Schuitemaker, I.B.M., 1988, *Cultuurtechnisch Vademecum*, Uitgave Cultuurtechnische Vereniging, Utrecht, ISBN 90-9002366-6.

Halcrow & Partners, DHI, 1993, River training studies of the Brahmaputra river, Dhaka, Bangladesh.

Haskoning, 2002, River bank erosion and bank protection review – Chandpur Town, Lower Meghna, Mission Report, Royal Haskoning, Project 9M083521, Nijmegen, the Netherlands.

Haskoning, 2006, Kribben van de Toekomst, kennisdocument over kribben in binnen- en buitenland, Royal Haskoning, reference 9R8485.A0/R0001/WDJO/MJANS/Nijm, Nijmegen, the Netherlands.

Hawkswood, M., & King, M., 2016, Bridge scour protection, *Proceedings of the ICSE-8*, London.

HEC-18, 2012, *Evaluating Scour at Bridges* (5th edition), Hydraulic Engineering Circular No. 18, Publication no FHWA-HIF-12-003, U.S.

HEC-23, 2009, *Bridge Scour and Stream Instability Countermeasures: Experience, Selection, and Design Guidance* (3rd edition), Hydraulic Engineering Circular No. 23 FHWA-NHI-09-111.

Herbich, J.B., 1991, Scour around pipelines, piles and sea walls, in: *Handbook of Coastal and Ocean Engineering*, 2, pp. 861–958, Gulf Publishing Company, Houston.

Hewlett, H.W.M., Boorman, L.A., & Bramley, M.E., 1987, Design of reinforced grass waterways, construction industry research and information association, Report No. 116, London.

Hjorth, P., 1975, *Studies on the Nature of Local Scour*, Department of Water Resources Engineering, Bulletin, Series A, 46, Lund Institute of Technology, University of Lund, Lund.

Hjulström, F., 1935, The morphological activity of rivers as illustrated by River Fyris, PhD thesis, Bulletin Geological Institute, Uppsala 25, Ch. III.

Hoan, N.T., 2008, Stone stability under non-uniform flow, PhD thesis, Delft University of Technology, Delft.

Hoeve, J.P., 2015, Rapid assessment tool for turbulence in backward facing step flow, MSc thesis, University of Twente.

Hoeve, J.B., Voortman, H.G., Ribberink, J.S., Warmink, J.J., & Uijttewaal, W.S.J., 2018, Rapid assessment of turbulence for feasibility design of hydraulic structures, River Flow2018, IAHR Committee on Fluvial Hydraulics, Lyon-Villeurbanne, France, September 5–8.

Hoffmans, G.J.C.M., 1990, Concentration and flow velocity measurements in a local scour hole, Report 4-90, Faculty of Civil Engineering, Hydraulic and Geotechnical Engineering Division, Delft University of Technology, Delft.

Hoffmans, G.J.C.M., 1992, Two-dimensional mathematical modelling of local-scour holes, Doctoral thesis, Faculty of Civil Engineering, Hydraulic and Geotechnical Engineering Division, Delft University of Technology, Delft.

Hoffmans, G.J.C.M., 1993, A hydraulic and morphological criterion for upstream slopes in local-scour holes, Report W-DWW-93-255, Ministry of Transport, Public Works and Water Management, Road and Hydraulic Engineering Division, Delft.

Hoffmans, G.J.C.M., 1994, Scour due to submerged jets (in Dutch), Report W-DWW-94-303, Ministry of Transport, Public Works and Water Management, Road and Hydraulic Engineering Division, Delft.

Hoffmans, G.J.C.M., 1995, Ontgrondingen rondom brugpijlers en aan de kop van kribben, Report W-DWW-94-312, Ministry of Transport, Public Works and Water Management, Road and Hydraulic Engineering Division, Delft, the Netherlands.

Hoffmans, G.J.C.M., 2009, Closure problem to jet scour, *Journal of Hydraulic Research*, 47(1): 100–109.

Hoffmans, G.J.C.M., 2010, Stability of stones under uniform flow, *Journal of Hydraulic Engineering*, 136(2) (February 1, 2010 and Hoffmans 2012).

Hoffmans, G.J.C.M., 2012, *The Influence of Turbulence on Soil Erosion*, Deltares Select Series 10, ISBN 978 90 5972 682 6, Eburon, Delft.

Hoffmans, G.J.C.M., & Booij, R., 1993a, The influence of upstream turbulence on local- scour holes, *Proceedings of the 25th IAHR-Congress, Tokyo* (also Ministry of Transport, Public

Works and Water Management, Road and Hydraulic Engineering Division, Delft, Report W-DWW-93–251).

Hoffmans, G.J.C.M., & Booij, R., 1993b, Two-dimensional mathematical modelling of local-scour holes, *Journal of Hydraulic Research, IAHR*, 31(5): 615–634.

Hoffmans, G.J.C.M., & Pilarczyk, K.W., 1995, Local scour downstream of hydraulic structures, *Journal of Hydraulic Engineering, ASCE*, 121(4): 326–340.

Hoffmans, G.J.C.M., & Verheij, H.J., 1997, *Scour Manual*. Balkema Publishers, Rotterdam, ISBN 90-800356-2-9.

Hoffmans, G.J.C.M., & Verheij, H.J., 2011, Jet scour, *Proceedings of the ICE – Maritime Engineering*, 164(4): 185–193.

Hoffmans, G.J.C.M., Hardeman, B., & Verheij, H.J., 2014, Erosion of grass covers at transitions and objects on dikes, *Proceedings of the ICSE-7*, Perth.

Hoffmans, G.J.C.M., & Buschman, F., 2018, Turbulence approach for predicting scour at abutments, *Proceedings of the ICSE-9*, Taipei, Taiwan.

Hofland, B., 2005, Rock and Roll, Turbulence-induced damage to granular bed protections, Doctoral thesis, Delft University of Technology, Delft.

Hopkins, G.R., Vance, R.W., & Kasraie, B., 1980, Scour around bridge piers, Report FHWA-RD-79-103, Federal Highway Administration, Offices of Research & Development Environmental Division, Washington, D.C.

Huismans, Y., van Velzen, G., O'Mahoney, T.S.D., Hoffmans, G., & Wiersma, A.P., 2016, Scour hole development in river beds with mixed sand-clay-peat stratigraphy, *Proceedings of the ICSE-8*, London.

ICOLD, 2015, Internal erosion of existing dams, levees and dikes, and their foundations, Volume 1: Internal erosion processes and engineering assessment. International Commission on Large Dams, Paris. Final preprint 19 February 2015 from: http://www.icoldcigb.org

Inglis, C.C., 1949, The behaviour and control of rivers and canals, Research Publication 13, C.W.I.N.R.S. Poona.

Izbash, S.V., 1932, *Construction of Dams by Dumping Stones into Flowing Water – Design and Practice*, Transactions of the Scientific Research Institute of Hydrotechnics, Leningrad, Vol. XVII, 12–65 (translated by A. Dovjikov. Corps of Engineers, U. S. Army, Eastport District, Eastport, Maine, 1935).

Izbash, S.V., & Khaldre, K.Y., 1970, *Hydraulics of River Channel Closures*, Butterworths, London.

Jacobsen, N.G., Van Velzen, G., & Fredsøe, J., 2014, Analysis of pile scour and associated hydrodynamic forces using proper orthogonal decomposition, 7th International Conference of Scour and Erosion, Perth, Australia.

Jansen, P.Ph. (Edt), 1979, *Principles of river engineering, the non-tidal alluvial river*, Pitman, London.

Johnson, P.A., 1992, Reliability-based pier scour engineering, *Journal of Hydraulic Engineering, ASCE*, 118(10): 1344–1358.

Johnson, P.A., 1995, Comparison of pier-scour equations using field data, *Journal of Hydraulic Engineering, ASCE*, 121(8): 626–629.

Jongeling, T.H.G., Blom, T., Jagers, A., Stolker, C., & Verheij, H., 2003, Design method granular protections, Delft Hydraulics, Technical Report Q2933/Q3018, Delft (in Dutch).

Jongeling, T.H.G., Jagers, H.R.A., & Stolker, C., 2006, Design of granular bed protections using a RANS 3D-flow model, *ISSMGE Proceedings of the 3rd International Conference on Scour and Erosion*, Amsterdam.

Jorissen, R.E., & Vrijling, J.K., 1989, Local scour downstream of hydraulic constructions, *Proceedings 23rd IAHR-congress, Ottawa*, B433–B440.

Kalinske, A.A., 1947, Movement of sediment as bed load in rivers, *Transactions, American Geophysical Union*, 28(4): 615–620.

Karelle, Y., et al., 2016, A full scale propeller wash erosion test on heterogeneous cohesive material, 8th International Conference on Scour and Erosion, London.

Khosronejad, A., Kang, S., & Sotiropoulos, F., 2012, Experimental and computational investigation of local scour around bridge piers. *Advance Water Resources Research*, 37: 73–85. http://dx.doi.org/10.1016/j.advwatres.2011.09.013.

Kidanemariam, A.G., & Uhlmann, M., 2014, Interface-resolved direct numerical simulation of the erosion of a sediment bed sheared by laminar channel flow, *International Journal of Multiphase Flow*, 67: 174–188. 10.1016/j.ijmultiphaseflow.2014.08.008.

Kim, H.S., Nabi, M., Kimura, I., & Shimizu, Y., 2014, Numerical investigation of local scour at two adjacent cylinders, *Advance Water Resources Research*, 70: 131–147, http://dx.doi.org/10.1016/j.advwatres.2014.04.018.

Kim, H.S., Nabi, M., Kimura, I., & Shimizu, Y., 2015, Computational modeling of flow and morphodynamics through rigid-emergent vegetation. *Advance Water Resources Research*, 84: 64–86, https://doi.org/10.1016/j.advwatres.2015.07.020.

Kirk, K.J., 2001, Instability of blanket bog slopes on Cuilcagh Mountain, N.W. Ireland, PhD thesis, University of Huddersfield.

Klaassen, G.J.R., & Vermeer, K., 1988, Confluence scour in large braided rivers with fine bed material, International Conference on Fluvial Hydraulics, Budapest, Hungary.

Klaassen, G.J., van Duivendijk, H., & Sarker, M.H., 2012, Performance review of Jamuna bridge river training works 1997–2009. *River Flow*, I: 3–16.

Konter, J.L.M., 1976, Stormvloedkering Oosterschelde, Werkgroep 8, Vormgeving en constructie sluitgaten, WL 8–44, Ontgrondingen bij de putten van de pijleroplossing, Report M1385, Delft Hydraulics, Delft.

Konter, J.L.M., 1982, Texaco Mittleplate platforms, Scour and bed protection, Report M1841, Delft Hydraulics, Delft.

Konter, J.L.M., & van der Meulen, T., 1986, Influence of upstream sand transport on local scour, Symposium on *Scale Effects* in *Modelling Sediment Transport Phenomena*, IAHR, Toronto.

Konter, J.L.M., & Jorissen, R.E., 1989, Prediction of time development of local scour, Hydraulic *Engineering Conference*, ASCE, New Orleans, Louisiana.

Konter, J.L.M., Jorissen, R.E., & Klatter, H.E., 1992, *Afsluitdammen regels voor het ontwerp*, Ministry of Transport, Public Works and Water Management, Road and Hydraulic Engineering Division, Delft.

Koote, M., 2017, A new physics based method for estimating the excess turbulence downstream of a structure, MSc thesis, University of Twente.

Launder, B.E., & Spalding, D.B., 1972, *Mathematical Models of Turbulence*, Academic Press, London.

Laursen, E.M., & Toch, A., 1956, *Scour Around Bridge Piers and Abutments*, Bulletin 4, Iowa Highway Research Board, State University of Iowa, Iowa Institute of Hydraulic Research, Iowa City.

Laursen, E.M., 1960, Scour at bridge crossings, *Journal of Hydraulic* Division, Proceedings of the American Society of Civil Engineers, 86(2): 39–54.

Lindenberg, J., & Koning, H.L., 1981, Critical density of sand, *Geotechnique*, 31(2): 231–245.

Liu, H.K., Chang, F.M., & Skinner, M.M. 1961, Effect of bridge construction on scour and backwater, Colorado State University, Department of Civil Engineering, Report CER60-HKL22.

Lu, S.S., & Willmarth, W.W., 1973, Measurements of the structure of the Reynolds stress in a turbulent boundary layer, *Journal of Fluid Mechanics*, 60(3): 481–511.

Mason, P.J., & Arumugam, K., 1985, Free jet scour below dams and flip buckets, *Journal of Hydraulic Engineering, ASCE*, 111(2): 220–235.

Mastbergen, D.R., & Berg, J.H. van den, 2003, Breaching in fine sands and the generation of sustained turbidity currents in submarine canyons. *Sedimentology*, 50: 635–637.

May, R.W.P., Ackers, J.C., & Kirby, A.M., 2002, Manual on scour at bridges and other hydraulic structures. CIRIA Report C551, ISBN 0 86017 551 0.

Melchers, R.E., 2002, *Structural Reliability Analysis and Prediction*, John Wiley & Sons, Hoboken NJ, ISBN 047-1983-241

Melville, B.W., & Coleman, S.E., 2000, *Bridge Scour*. Water Resources Publications, Colorado, ISBN 1-887 201-18-1.

Mesbahi, J., 1992, On combined scour near Groynes in river bends, Delft Hydraulics (also UNESCO/IHE MSc thesis), Delft.

Mirtskhoulava, Ts.Ye., 1988, *Basic Physics and Mechanics of Channel Erosion*, Gidrometeoizdat, Leningrad.

Mirtskhoulava, Ts.Ye., 1991, Scouring by flowing water of cohesive and non-cohesive beds, *Journal of Hydraulic Research, IAHR*, 29(3): 341–354.

Monaghan, J.J., 1992, Smoothed particle hydrodynamics, *Annual Review of Astronomy and Astrophysics*, 30: 543–574, https://doi.org/10.1146/annurev.aa.30.090192.002551

Mosselman, E., Shishikura, T., & Klaassen, G.J., 2000, Effect of bank stabilization on bend scour in anabranches of braided rivers, *Physics and Chemistry of the Earth, Part B*, 25(7–8): 699–704.

Mosonyi, E., & Schoppmann, B., 1968, Ein Beitrag zur Erforschung von örtlichen Auskolkungen hinter geneigten Befestigungsstrecken in Abhängigkeit der Zeit, Mitteilungen des Theodor-Rehbock-Flußbaulaboratoriums, Universität Karlsruhe.

Nabi, M., De Vriend, H.J., Mosselman, E., Sloff, C.J., & Shimizu, Y., 2012, Detailed simulation of morphodynamics: 1. Hyrodynamic model, *Water Resources Research*, 48: W12523. http://dx.doi.org/10.1029/2012WR011911.

Nabi, M., De Vriend, H.J., Mosselman, E., Sloff, C.J., & Shimizu, Y., 2013a, Detailed simulation of morphodynamics: 2. Sediment pick-up, transport and deposition, *Water Resources Research*, 49: 4775–4791. http://dx.doi.org/10.1002/wrcr.20303.

Nabi, M., De Vriend, H.J., Mosselman, E., Sloff, C.J., & Shimizu, Y., 2013b, Detailed simulation of morphodynamics: 3. Ripple and dunes, *Water Resources Research*, 49: 5930–5943. http://dx.doi.org/10.1002/wrcr.20457.

Nakagawa, H., & K. Suzuki, 1976, Local scour around bridge pier in tidal current, *Coastal Engineering in Japan*, 19: 89–100.

NCHRP, 2007, Countermeasure Concepts and Criteria, NCHRP, Report 587 National Co-operative Highway Research Program, 2007.

NCHRP, 2011, *Evaluation of Bridge Scour Research: Pier Scour Processes and Predictions*, Robert Ettema, George Constantinescu, Bruce Melville

Neill, C.R., 1968, *A Re-examination of the Beginning of Movement for Coarse Granular Bed Materials*, Hydraulic Research Station, Wallingford.

Neill, C.R., 1973, *Guide to Bridge Hydraulics*, University of Toronto Press, Toronto.

NEN 3651, 2012, *Aanvullende eisen voor buisleidingen in of nabij waterstaatswerken*, NEN, Delft

Nezu, I., 1977, *Turbulent Structure in Open-Channel Flows* (Translation of Doctoral Dissertation in Japanese), Kyoto University, Kyoto.

Nezu, I., & Nakagawa, H., 1993, *Turbulence in Open-Channel Flows*, Balkema, Rotterdam.

Nezu, I., & Tominaga, A., 1994, Response of velocity and turbulence to abrupt changes from smooth to rough beds in open-channel flows, *Proceedings of Symposium, Fundamentals and Advancements in Hydraulic Measurements and Experimentation*, ASCE, Buffalo, NY, 195–204.

Nicoud, F., 2007, *Unsteady Flows Modelling and Computation*, University Montpellier II and I3M-CNRS UMR 5149, France (franck.nicoud@univ-montp2.fr)

Odgaard, A.J., 1981, Transverse slope in alluvial channel bends, *Journal of Hydraulic Engineering, ASCE*, 107(HY12): 1677–1694.

O'Mahoney, T., 2018, Flow field weir Grave during calamity (in Dutch). Technical Report, Deltares, Delft.

Osman, A.M., & Thorne, C.R., 1988, Riverbank stability analysis (I : theory and II : Applications). *Journal of Hydraulic Engineering*, 114(2): 134–172.

Paintal, A.S., 1971, Concept of critical shear stress in loose boundary open channels, *Journal of Hydraulic Research*, 9(1): 91–113.

PIANC, 2015, Guidelines for protecting berthing structures from scour by ships. PIANC Report 180, Brussels.

Pilarczyk, K.W., 1984, Interaction water motion and closing elements, in: *The Closure of Tidal Basins*: 387–405, Ed. Board: Huis in 't Veld, J.C. et al, Delft University Press, Delft.

Pilarczyk, K.W., 2003, Design of revetments. Dutch Public Works Department, Report.

Przedwojski, B., Blazejewski, R., & Pilarczyk, K.W., 1995, *River Training Techniques: Fundamentals, Design and Applications*, Balkema, Rotterdam.

Qayoum, A., 1960, Die Gesetzmä igkeit der Kolkbildung hinter unterströmten Wehren unter spezieller Berücksichtigung der Gestaltung eines beweglichen Sturzbettes, Dissertation, Technischen Hochschule Carolo-Wilhelmina, Braunschweig, Germany.

Raaijmakers, T.C., van Velzen, G., Hoffmans, G.J.C.M., Bijlsma, A., Verbruggen, W., & Stronk, J., 2012, Stormvloedkering Oosterschelde: ontwikkeling ontgrondingskuilen en stabiliteit bodembescherming. Rapport 1206907-003, Deltares, Delft.

Rajaratnam, N., 1976, *Turbulent Jets*, Elsevier, Amsterdam.

Rajaratnam, N., 1981, Erosion by plane turbulent jets, *Journal of Hydraulic Research, IAHR*, 19(4): 339–358.

Rajaratnam, N., & Berry, B., 1977, Erosion by circular turbulent wall jets, *Journal of Hydraulic Research, IAHR*, 15(3): 277–289.

Raudkivi, A.J., 1993, *Sedimentation: Exclusion and Removal of Sediment from Diverted Water*, Balkema, Rotterdam.

Raudkivi, A.J., & Ettema, R., 1985, Scour at cylindrical bridge piers in armoured beds, *Journal of Hydraulic Engineering, ASCE*, 111(4): 713–731.

Richardson, E.V., Simons, D.B., & Julien, P., 1990, *Highways in the River Environment*, FHWA-HI-90-016, Federal Highway Administration, US. Department of Transportation, Washington, D.C.

RIZA, 2005a, Ontgrondingskuilen in de Waal, RIZA werkdocument 2005.082x

RIZA, 2005b, Inschatting van de kans op achterloopshcid bij kribben, RIZA werkdocument 2005.148X.

Römisch, K., & Schmidt, E., 2012, Durch Schiffspropeller verursachte Kolkbildung, *HANSA International Maritime Journal*, 1012(9): 171–175.

Roubos, A.A., Blokland, T., & van der Plas, A.F., 2014, Field tests propeller scour along quay walls, PIANC congress San Francisco, USA.

Ruff, J.F., Abt, S.R., Mendoza, C., Shaikh, A., & Kloberdanz, R., 1982, Scour at culvert outlets in mixed bed materials, Report FHWA/RD-82/011, Colorado State University, Fort Collins, Colorado.

Sagaut, P., Deck, D., & Terracol, M., 2006, *Multiscale and Multiresolution Approaches in Turbulence*, Imperial College Press, London.

Schiereck, G.J., & Verhagen, H.J., 2012, *Introduction to Bed, Bank and Shore Protection*, VSSD, Delft.

Schlichting, H., 1951, *Grenzschichttheorie*, Braunsche Hofbuchdruckerei, Karlsruhe.

Schoklitsch, A., 1932, Kolkbildung unter Überfallstrahlen, *Die Wasserwirtschaft*, 24: 341–343.

Schoklitsch, A., 1935, *Stauraumverlandung und Kolkabwehr*, Springer, Vienna.

Schoklitsch, A., 1962, Die Kolkbildung, *Handbuch des Wasserbaues*, 1: 200–210.

Shafii, I., Briaud, J.L., Chen, H.C., & Shidlovskaya, A., 2016, Relationship between soil erodibility and engineering properties, *Proceedings of the ICSE-8*, London.

Shalash, M.S.E., 1959, Die Kolkbildung beim Ausflu unter Schützen, Dissertation, Technischen Hochschule München, Germany.

Sheppard, D.M., & Glasser, T.L., 2004, *Sediment Scour at Piers with Complex Geometries* (2nd edition), ICSE, Singapore.

Shields, A., 1936, Anwendung der Ähnlichkeitsmechanik und der Turbulenzforschung auf die Geschiebebewegung, Mitteilungen Preussischen Versuchsanstalt für Wasserbau und Schiffbau, 26, Berlin.

Silvis, F., 1988, Oriënterende studie naar grondmechanische aspecten bij ontgrondings- kuilen, Report CO-291720/12, Grondmechanica Delft, Delft.

Sloff, C.J., van Spyk, A., Stouthamer, E., & Sieben, A., 2013, Understanding and managing the morphology of branches incising into sand-clay deposits in the Dutch Rhine Delta, *International Journal of Sediment Research*, 28(2), 127–138.

Smith, W.L., 1999, Local structure-induced sediment scour at pile groups, MS thesis, UFL/COEL-99/003, University of Florida, Gainesville, FL.

Sonneville, B., Joustra, R., & Verheij, H., 2014, Winnowing at circular piers under currents, 7th ICSE, Perth.

Sowers, G.F., 1979, Introductory soil mechanics and foundations, Geotechnical Engineering, Macmillan Publishers

Stoutjesdijk, T., de Kleine, M., de Ronde, J., & Raaijmakers, T., 2012, Stormvloedkering Oosterschelde: ontwikkeling ontgrondingskuilen en stabiliteit bodembescherming, Hoofdrapport, Deltares-Rapport 1206907-005.

Straathof, J., 2015, *Bodembescherming Rond Brugpijlers*, Afstudeerscriptie, Hogeschool van Amsterdam, Amsterdam.

Struiksma, N., Olesen, K.W., Flokstra, C., & de Vriend, H.J., 1985, Bed deformation in curved alluvial channels, *Journal of Hydraulic Research, IAHR*, 23(1): 57–79.

Struiksma, N., & Verheij, H., 1995, Bend scour, Personal Communications (Unpublished).

TAW, 2003, Leidraad Kunstwerken, Technische Adviescommissie Waterkeringen, document DWW-2003-059, ISBN 90-369-5544-0, Delft

ten Brinke, W.B.M., Kruyt, N.M., Kroon, A., & Van Den Berg, J.H., 1999, *Erosion of Sediments between Groynes in the River Waal as a Result of Navigation Traffic*: 147–160, Blackwell Publishing Ltd., Hoboken NJ.

ten Brinke, W.B.M., 2003, De sedimenthuishouding van kribvakken langs de Waal: het langjarig gedrag van kribvak stranden, de invloed van scheepsgeïnduceerde waterbeweging en morfologische processen bij hoge en lage afvoeren. Min. van V & W, RWS, RIZA Rapport 2003.002, ISBN 9036954827, Lelystad.

ten Brinke, W.B.M., Schulze, F.H., & van Der Veer, P., 2004, Sand exchange between groyne field beaches and the navigation channel of the Dutch Rhine: the impact of navigation versus river flow, *River Research and Applications*, 20(8): 899–928.

Thorne, C.R., 1993, Prediction of near-bank velocity and scour depth in meander bends for design of riprap revetments, in: River, coastal and shoreline protection: erosion control using riprap and armour stone, Riprap Workshop, Fort Collins, Colorado: 980–1007.

Tsujimoto, T., 1988, Analytical approach to some practical aspects of local scour around bridge piers, In: P.A. Kolkman, J. Lindenberg, & K.W. Pilarczyk (eds), *Modelling Soil-Water-Structure Interactions (SOWAS 88), Proceedings of the International Symposium, Delft*: 137–145, Balkema, Rotterdam.

US Bureau of Reclamation, 1955, Research studies on stilling basins, energy dissi-pators, and associated appurtenances, Hydraulic Laboratory Report Hyd-399.

van den Ham, G.A., de Groot, M.B., & Mastbergen, D.R., 2014, A semi-empirical method to assess flow-slide probability. In: Krastel et al. (eds), *Proceedings of the 6th International Symposium Submarine Mass Movements and Their Consequences*, 213–223, Springer International Publishing, Swiss.

van der Meulen, T., & Vinjé, J.J., 1975, Three-dimensional local scour in noncohesive sediments, *Proceedings of the 16th LAHR-Congress, Sao Paulo, Brasil* (also Delft Hydraulics, Delft, Publication 180).

van der Wal, M., 1991, Investigation to scour near a horizontal constriction, Report Q935. Delft Hydraulics, Delft.

van Mierlo, M.C.L.M., & de Ruiter, J.C.C., 1988, Turbulence measurements above dunes; Report Q789, vol. 1 and 2, Deltares, Delft.

van Rijn, L.C., 1982, Equivalent roughness of alluvial bed, *Journal of Hydraulic Engineering, ASCE*, 110(10): 1431–1456.

van Rijn, L.C., 1984, Sediment transport, Part 1: Bed load transport, *Journal of Hydraulic Engineering*, 110(10): 1431–1456; Part 2: Suspended load transport, 110(11): 1613–1641; Part 3: Bed forms and alluvial roughness, 110(12): 1733–1754.

van Rijn, L.C., 1985, Mathematical modelling for sediment concentration profiles in steady flow, In W. Bechteler (ed.), *Transport of Suspended Solids in Open Channels, Proceedings of the Euromech 192, Munich/Neubiberg*, Balkema, Rotterdam (also Delft Hydraulics, Delft, Publication 354).

van Rijn, L.C., 1993, *Principles of Sediment Transport in Rivers, Estuaries and Coastal Seas*, Aqua Publications, Amsterdam.

Van Rijn, L.C., 2007, *Manual Sediment Transport Measurements in Rivers, Estuaries and Coastal Seas*, Aqua Publcations, Amsterdam.

van Velzen, G., Raaijmakers, T.C., & Hoffmans, G.J.C.M., 2014, Scour development around the Eastern Scheldt storm surge barrier-field measurements and model predictions, ICSE-7, Perth, Australia.

Vanoni, V.A., 1966, Sediment transportation mechanics: Initiation of motion, (Progress Report of Task Committee), *Journal of Hydraulics Division, ASCE*, 92(HY2): 291–314.

Vanoni, V.A. (ed.), 1977, *Sedimentation Engineering: Manual 54*, ASCE, New York, ISBN 978-0-7844-7134-0.

Veiga da Cunha, L., 1971, Erosoes lokalizadas junta de obstáculos salientes de margens (Local scour at obstacles protruding from river banks), Thesis presented to the University of Lisbon, LNEC, Lisboa, Portugal.

Veronese, A., 1937, Erosioni di fondo a valle di uno scarico, *Annali dei Lavori Pubblici*, 75(9), 717–726.

Vrijling, J.K., 1990, Probabilistic design of flood defences, In: K.W. Pilarczyk (ed.), *Coastal Protection, Proceedings of the Short Course, Delft*: 39–99, Balkema, Rotterdam.

Vrolijk, J.W.H., Kruyt, C., & Van der Wal, M., 2005, Innovative abutments, Prototype experiment Haaften. T0-measurement [in Dutch], Draft Report, Oranjewoud, Heerenveen, The Netherlands.

VTV, 2016, Voorschrift Toetsen op Veiligheid, Technisch Deel (WTI2017), Rapport Deltares, 1220078-000, Delft.

Whittaker, J.G., & Schleiss, A., 1984, Scour related to energy dissipators for high head structures, Mitteilungen der Versuchsanstalt für Wasserbau, Hydrologie und Glaziologie, 73, Technischen Hochschule Zürich, Switzerland.

Wilderom, M.H., 1979, Resultaten van het vooronderzoek langs de Zeeuwse stromen, Report 75.2, Rijkswaterstaat, Directie Waterhuishouding en Waterbeweging Studiedienst Vlissingen.

Winterwerp, J.C., 1989, *Cohesive Sediments, Flow Induced Erosion of Cohesive Beds*, Rijkswaterstaat/Delft Hydraulics, Delft.

Winterwerp, J.C., van Kesteren, W.G.M., van Prooijen, B., & Jacobs, W., 2012, A conceptual framework for shear flow-induced erosion of soft cohesive sediment beds, *Journal of Geophysical Research*, 117, C10020.

Yalin, M.S., 1972, *Mechanics of Sediment Transport*, Pergamon, New York.

Yalin, M.S., 1992, *River Mechanics*, Pergamon, New York.

Yerroa, A., Rohe, A., & Sogac, K., 2017, Modelling internal erosion with the material point method, *Conference: 1st International Conference on the Material Point Method*, MPM, Delft, the Netherlands.

Yossef, M.F.M., 2002, The effects of groynes on river, a literature review, Technical Report, Delft University.

Zanke, U., 1978, Zusammenhänge zwischen Strömung und Sedimenttransport, Teil 1: Berechnung des Sedimenttransportes, -allgemeiner Fall-, Teil 2: Berechnung des Sedimenttransportes hinter befestigten Sohlenstrecken, -Sonderfall zweidimensionaler Kolk-, Mitteilungen des Franzius-Instituts der TU Hannover, Heft 47, 48.

Zanke, U., 1994, Kolkschutz durch integrierte Schütteinrichtungen, *Wasser und Boden*, 2: 20–22.

Zhang, X., Chen, Z., & Liu, Y., 2016, *The Material Point Method, A Continuum-Based Particle Method for Extreme Loading Cases*, eBook ISBN: 9780124078550, Hardcover ISBN: 9780124077164, Academic Press, Cambridge, MA.